Supervised Machine Learning

Supervised Machine Learning
Optimization Framework and Applications with SAS and R

Tanya Kolosova PhD
Associates in Analytics Inc., Boca Raton, Florida

Samuel Berestizhevsky MSc
Associates in Analytics Inc., Boca Raton, Florida

CRC Press
Taylor & Francis Group
Boca Raton London New York

CRC Press is an imprint of the
Taylor & Francis Group, an **informa** business
A CHAPMAN & HALL BOOK

First edition published 2021
by CRC Press
6000 Broken Sound Parkway NW, Suite 300, Boca Raton, FL 33487-2742

and by CRC Press
2 Park Square, Milton Park, Abingdon, Oxon, OX14 4RN

First issued in paperback 2022

Publisher's Note
The publisher has gone to great lengths to ensure the quality of this reprint but points out that some imperfections in the original copies may be apparent.

Visit the Taylor & Francis Web site at
http://www.taylorandfrancis.com

and the CRC Press Web site at
http://www.crcpress.com

ISBN: 978-0-367-27732-1 (hbk)
ISBN: 978-0-367-53882-8 (pbk)
ISBN: 978-0-429-29759-5 (ebk)

DOI: 10.1201/9780429297595

Typeset in Palatino
by codeMantra

Dedication

To our children—love and meaning of our lives.

Contents

Part III

Acknowledgments

Writing this book was a real adventure. It summarizes our experience over many years of solving problems related to applying machine learning methods to real-life data. During these years, we met many people who, by sharing their experience, asking questions, and challenging our solutions, helped to crystalize this book's ideas. Among them, Samuel would like to thank Maurice "Hank" Greenberg for being a unique inspiration and Charles Dangelo for his expert advice in the insurance domain. Tanya is grateful for the knowledge of the design of experiments shared by Prof. David M. Steinberg from Tel Aviv University, Israel, and Prof. Carl Schwarz from Simon Fraser University, Canada. We want to thank the SAS Institute as our long-term software of choice, as well as the R Core Team and contributors.

We would also like to thank everyone on the Taylor and Francis/CRC Press team for their support and assistance in preparing this book. Special thanks to David Grubbs, the acquisition editor, for his great support. Thanks to Rebecca, Sofia, Varun, and other editorial staff members for helping this book come to fruition.

Special thanks to our daughter Efrat for her performances of Chopin and Debussy's piano works that provided a calming and inspirational atmosphere.

Authors

Tanya Kolosova is a statistician, software engineer, educator, and co-author of two books on statistical analysis and metadata-based applications development using SAS. She is an actionable analytics expert and has extensive knowledge of software development methods and technologies, artificial intelligence methods and algorithms, and statistically designed experiments.

Samuel Berestizhevsky is a statistician, researcher, and software engineer. Together with Tanya, Samuel co-authored two books on statistical analysis and metadata-based applications development using SAS. Samuel is an innovator and expert in the area of automated actionable analytics and artificial intelligence solutions. His extensive knowledge of software development methods, technologies, and algorithms allows him to develop solutions on the cutting edge of science.

Introduction: Challenges in the Application of Machine Learning Classification Methods

According to Bishop (2006), machine learning (ML) is a scientific study of statistical models and algorithms to help a computing system accomplish a specific task without using explicit instructions, but relying on patterns and inference instead. Such patterns and inferences can be extracted from sample data, also called "training datasets" in the machine learning domain, using machine learning algorithms.

ML is a subset of artificial intelligence technology. It automatically learns and improves the performance with the pace of time, interactions, and experiences. Building initial machine learning models, or classifiers, is an iterative process, where we start with initial hypotheses about which data can be useful and how to structure different sets of features, which machine learning methods can learn better from our data, and how to "tune" multiple hyperparameters to achieve robust classification results. The machine learning development process becomes a nightmare if it does not use a framework for easy, reliable, meaningful, verifiable, and reproducible development.

A large amount of available ML methods creates a challenge for researchers: how to develop, verify, reproduce, and compare models using different ML methods? Utilizing the same or similar data for modeling by different methods naturally leads to outcomes that differ in their form or quality or both. We want to be able to compare results from different models and to draw conclusions about the utility of using one model over the other. The design of statistical experiments may work as a framework that helps us to systematically compare different models. Such experiments, which include determining which class of models to use and what types of features to include, produce a number of different results.

Reproducibility is the ability to reproduce the analyses.

Comparability is the ability to compare results produced by different modeling approaches.

In this book, we describe an AI framework that helps developers to achieve reproducibility and comparability of machine learning experiments and to define ML development processes in precise, transparent, easily changeable, and verifiable ways.

Fundamentals of Supervised Classification

Machine learning can be roughly categorized into three types: supervised learning, unsupervised learning, and reinforcement learning. In this book, we only deal with supervised machine learning. The most common supervised machine learning task is classification.

Supervised classification can be described as the task of automatically assigning objects to their respective classes on the basis of numerical measurements derived from these objects. Classifiers are the tools that implement the actual classification from these measurements to the so-called class labels. The field of supervised classification studies ways of constructing such classifiers. The main idea behind supervised learning methods is that of learning from examples: given several so-called training datasets that describe input–output relations, to what extent can the general mapping be learned that takes any new and unseen objects to their correct classes?

The basic problem of machine learning is probably the following. We have a training dataset of cases

$$\left(x_1, y_1\right), ..., \left(x_n, y_n\right)$$

where each case $\left(x_i, y_i\right)$, $i = 1, ..., n$ consists of an object x_i (a vector of features) and its label y_i. The problem is to predict the label y_{n+1} of a new object x_{n+1}. Usually, the goal of classification is to produce a prediction $\hat{y}_{n+1}$ that is likely to coincide with the true label y_{n+1}, and this goal should be complemented with some measure of its reliability. There is a clear tradeoff between accuracy and reliability: we can improve the former by relaxing the latter and vice versa.

Generalization Problem: Bias–Variance Tradeoff

The quality of a classifier is defined by its ability to correctly classify an input vector into an appropriate class and is measured by metrics derived from confusion, or misclassification, matrix. There is no such metric that can measure a classifier quality alone: at least a pair of metrics should be used. For example, sensitivity and specificity estimate true positive and true negative rates, respectively, and we are interested in classifiers with both of these metrics approaching 1.

The bias–variance tradeoff is the idea that finding a minimum-risk classification method involves striking a balance between minimizing bias (being right on average) and minimizing variance (being stable with respect to variation in training datasets). A classification method that performs poorly due to high variance is said to overfit when presented with data. One that performs poorly due to high bias is said to underfit.

Low bias means

- the average accuracy metric of classification results is close to the truth.

High bias means

- the average accuracy metric of classification results is far from the truth,
- the classification method is not sufficiently flexible, and
- individual results of classification accuracy metric are not adequately adapted to the data.

Low variance means

- each individual accuracy metric of classification is close to the average accuracy metric,
- individual accuracy metric tends to be similar to one another, and
- the classification method is "stable" with respect to which dataset (of all possible datasets) it is applied.

High variance means

- individual accuracy metrics are often far from the average accuracy,
- individual accuracy metrics are quite different from one another, and
- the classification method is very sensitive to which dataset it is applied.

The challenge of bias and variance estimation can actually be easily addressed by applying two readily available statistical methodologies:

- bootstrap and
- mixture experiments.

We demonstrate in this book how the application of m-out-of-n bootstrap allows creation of thousands of training datasets and in turn how the application of the ML method to these datasets allows us to estimate bias and variance. The statistical approach of the mixture experiment is applied to contaminate data and to find the level of contamination that "breaks" the classifier. This helps to identify robust classifiers that are better positioned for unseen data.

Challenges of Deployment

The useful life of the machine learning application begins when it is deployed to production. From this point, application modifications are often required. These modifications may consume a great deal of effort in recoding and even redesigning. Generally, the better the machine learning application, the longer its life. Modifications and maintenance of the application are usually performed by people other than the original implementers. This problem is very difficult to solve because the application internals cannot be easily understood. The code-free framework described in this book addresses these problems. Generally, the framework outlined in this book addresses the most fundamental problems of existing machine learning application development and deployment technologies: the failure to recognize users' true needs and the inability to develop machine learning solutions quickly to meet these needs.

Machine learning applications have all the challenges of the traditional software systems, plus an additional set of machine learning–specific issues – reproducibility and comparability. These two machine learning–specific issues can be solved by applying the principles of statistically designed experiments combined with bootstrap along with testing the robustness of the machine learning solution through data contamination processes (flipping or corruption of labels of training dataset). The principles and applications of designed experiments, bootstrap, and data contamination are discussed in this book.

Another challenge is the selection of the production classifier. If, for example, the criterion of a classifier quality is to maximize sensitivity and specificity, then we may encounter a difficult choice. Let us say, we have classifier A with a sensitivity of 0.85 and a specificity of 0.88, and classifier B with a sensitivity of 0.87 and a specificity of 0.87 – then which of them should we choose?

In this book, we want to explore a different approach, wherein we do not need to limit our choice by one classifier. Both classifiers, A and B, have a very good classification quality. More than that, we can find a tens of classifiers with very similar sensitivity and specificity. We also know that the sensitivity and specificity of each classifier will vary depending on the datasets they are applied to. So, instead of choosing one classifier, we will use all classifiers that have high-quality metrics, and the final decision on the classification result will be based on a voting mechanism using some information criteria. This approach will immediately improve the classification process.

Source Code

The source code for the book can be continuously improved and extended after the book has been published. The source code is located on GitHub:

https://github.com/smlof/Supervised-Machine-Learning--Optimization-Framework

Part I

Part I

1

Introduction to the AI Framework

The artificial intelligence (AI) framework should be algorithm and technology agnostic. This means that it should not focus on a specific machine learning methodology and specific systems or software, but it applies to the design, application, and use of AI in general.

In this book, we describe the AI framework that helps to create machine learning classifiers in such a way that they produce a sufficiently good classification of unseen data. The framework incorporates data dictionaries that define processes such as statistically designed experiments, bootstrapping, and data "contamination." This framework is fully transparent as all data processes are defined as metadata that ensures comparability and reproducibility of the results of the supervised classification.

Because the AI framework is fully independent of technology, it can be implemented using different software platforms. In this book, we present the implementation of the AI framework using two different software tools: SAS and R.

SAS is an integrated software system that enables accessing data across multiple sources, manipulating data, performing sophisticated analyses, and much more, and is one of the major choices of corporate business, though a bit costly.

R (R Core Team, 2018) is an open-source programming language developed by researchers, and many latest statistical and analytical methods are developed with R. This language has extensive documentation along with a supportive community and is widely used among statisticians and data scientists for developing statistical software and data analysis.

1.1 Components of the AI Framework and Their Interaction

The main idea of the AI framework proposed in this book is to create a systematic approach that resolves the challenges of machine learning modeling.

The process of building a machine learning classifier requires making choices that very often cannot be informed by intuition, understanding of a business problem, or pre-analysis of data. For example, making a decision about the feature subset requires considering different hypotheses, and it is reasonable to consider several different feature sets. Which one of them to choose eventually requires to go through the modeling process and comparison of classifiers built on different feature sets.

Choosing a machine learning method is challenging as well, and in many instances, the decision is made based on the convenience of the method and expected interpretability of the results, rather than based on theoretical and technical aspects of machine learning methods. If we do not want to limit ourselves to only one machine learning method, but rather want to try a few of them, we need to estimate and compare the classifiers built by those methods. Now we have to keep in mind that the quality of an estimated classifier depends not only on the machine learning method used, but on the feature set as well.

For each machine learning method, we need to assign values to the hyperparameters of that method, and it is obvious that the values of the hyperparameters impact the quality of the estimated classifier. Different methods have different numbers of hyperparameters, and finding an optimal set of values becomes a very difficult problem.

Figure 1.1 describes some most important components of the AI framework regarding which a decision should be made, but how? The statistical design of an experiment is a methodology that helps here. It allows us to create a plan, execute it, compare the results according to the selected metrics of the classifier quality, and repeat if needed. It often happens that we end up with several classifiers that have similar classification quality.

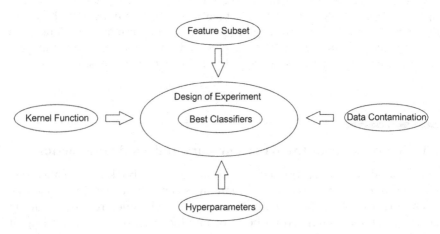

FIGURE 1.1
Components of the AI framework.

1.2 AI Framework in Detail

AI framework addresses the challenges of building classifiers. These challenges include not only the difficulties of making selection of feature sets, machine learning methods, and values of their hyperparameters based on the classifier quality, but also the problems that arise during the process of classifier building. These details are presented in Figure 1.2.

1.2.1 Creating Training and Test Datasets

As a first step, the AI framework deals with creating training and test datasets. The familiar approach of dividing available data into two parts, and being concerned only with a proportion of such a division, e.g., 50-50 and 70-30, has a built-in drawback. It relates to bias and dependency within data. It also relates to the situation that estimating a classifier on one dataset does not allow us to measure variance and bias of the classifier quality and does not allow us to evaluate the classifier robustness. In the developed AI framework, the bootstrap approach is used to address this drawback.

- Bootstrap is used to create multiple—and if data permits, hundreds or even thousands of—training and test datasets. It allows us to estimate and correct bias in data, increase data variability, and decrease dependency within data.

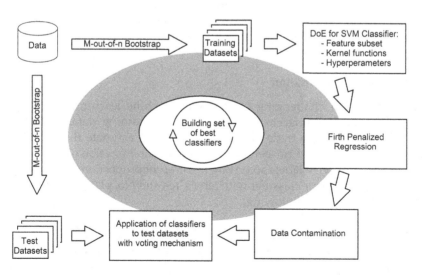

FIGURE 1.2
AI framework process.

- Estimation of a classifier on multiple training datasets produces a sample of the classifier quality metrics that now allows for a comprehensive estimation of the classifier quality.
- Use of multiple, instead of a single, test datasets helps to validate the ability of the classifier to deal with unseen data.

1.2.2 Design of Experiments for a Classifier

As a next step, the AI framework addresses the question of parameters selection, where the plan is created based on the statistical methodology of the design of experiments. The plan creates combinations of all parameters of interest in a way that the results can be analyzed and significant insights can be drawn. For example, we may consider five possible feature sets, the support vector machine method with three different kernels, and three intervals of values for each hyperparameter. The plan (design of experiments) will define how many different setups should be used to build a classifier and how many training datasets each setup should be applied to.

1.2.3 Firth Logistic Regression

Classifier results, by the nature of machine learning algorithms, do not have probabilistic characteristics. This means that if an outcome is predicted as 1, this prediction is not accompanied by a probability of it being correct. Being able to associate probabilities with predictions increases the quality of a classifier. Firth logistic regression is a statistical method that works very well with classifier prediction, adding to it a probability value.

1.2.4 Data Contamination

One of the most challenging problems of machine learning modeling is generalization. Building classifiers on multiple training datasets helps to address bias–variance trade-off, but it still cannot estimate the robustness of the classifier toward unseen data. Planned contamination of data means random corruption of labels according to different proportions. For example, in the case of a binary classifier with labels Yes and No, planned contamination may look like the following:

1. "Flip" 0.5% of Yes labels to No, and 0.5% of No labels to Yes,
2. "Flip" 0.5% of Yes labels to No, and 1% of No labels to Yes,
3. "Flip" 1% of Yes labels to No, and 0.5% of No labels to Yes,
4. Etc.

The idea is, essentially, to find at what level of contamination the classifier loses its quality significantly and to exclude those classifiers that lose their quality with just slight contamination.

1.2.5 Best Classifiers

As mentioned earlier, we do not limit classification solution to only one classifier. On the one hand, it is often impossible to choose the best classifier based on the pairs of metrics like sensitivity and specificity. It is a common situation that some classifiers are better on sensitivity, others are better on specificity, and yet others are not the best but very good on both. Essentially, we may consider a classifier as a good one when both metrics are above the thresholds established based on the need of the application.

On the other hand, we do not know how each classifier would perform on unseen data. Although the AI framework creates processes that help to improve the robustness of classifiers, it should be expected that the classification quality would decrease on new data. Thus, instead of one classifier, the AI framework uses a set of classifiers that made it to the top and then applies the voting approach to make a decision regarding the predicted classification.

1.3 SAS Procedures for the AI Framework Components

This book should not be considered as a programming tutorial for SAS or R. There is extensive literature dedicated to this topic. However, we would like to demonstrate how some of the capabilities can be used and we hope that such examples will allow the reader to extend the AI framework by using other capabilities of SAS, R, or additional software.

SAS has a powerful set of tools for machine learning methods, design of experiments (Lawson, 2010), analysis of experiment results, and much more. In this book, we use SAS procedures as elementary units for the AI framework. But this is not all. SAS has a powerful macro language that allows us to execute tasks according to the metadata stored in a data dictionary. This is the capability of SAS that we leverage in the AI framework (Guard, 2018).

1.4 R Libraries for the AI Framework Components

R, by the nature of it being an open-source language, grows continuously in the area of machine learning and has multiple packages that can be used for advanced machine learning projects. In this book, we do not make an attempt to use all of the packages, but we invite the reader to consider an opportunity to use different machine learning methods and R packages. The "e1071" package is one of the most widely used R packages for machine learning, and the one we leverage in the book. This package allows us to implement such methods as support vector machine (SVM), shortest path computation, bagged clustering, naive Bayes classifier, short-time Fourier transform, and fuzzy clustering. Another package "kernlab" also implements SVM, as well as kernel feature analysis, ranking algorithm, dot product primitives, Gaussian process, and many more. This package provides different kernel functions such as polynomial, hyperbolic tangent, and Laplacian. For a neural network, "nnet" package can be used. Although this package has a single-layer limitation, it is still one of the most popular packages for neural networks. Random forest classifier that generates multiple decision trees can be built using the "randomforest" packages for machine learning. Yet another R machine learning package is "mboost" with the functional gradient descent algorithm for optimizing general risk functions.

In addition to the machine learning capabilities, R offers the design of experiment packages that are necessary for the AI framework (Lalanne, 2006). For example, "doe.base" package offers utilities to design and analyze full factorial experiments and orthogonal arrays; "frf2" enables design and analysis of fractional factorial experiments with two-level factors; "dicedesign" allows us to design and execute computer experiments; "bsmd" is the package for Bayes screening and model discrimination utilities; and "doe.wrapper" is a wrapper package for design of experiment functionality.

A variety of functions and procedures developed in SAS and R are the building blocks for the AI framework.

References

Guard, R. 2018. *Machine Learning with SAS®: Special Collection*. Cary, NC: SAS Institute Inc.

Lalanne, C. 2006. *R Companion to Montgomery's Design and Analysis of Experiments*. https://www.semanticscholar.org/

Lawson, J. 2010. *Design and Analysis of Experiments with SAS*. Chapman and Hall: CRC.

R Core Team. 2018. *R: A Language and Environment for Statistical Computing*. Vienna, Austria: R Foundation for Statistical Computing.

2

Supervised Machine Learning and Its Deployment in SAS and R

2.1 Introduction

Machine learning is the study of methods that program computers to learn from data. The formal definition of machine learning algorithms was given by Tom Mitchell (1997), who stated that "A computer program is said to learn from experience E with respect to some class of tasks T and performance measure P if its performance at tasks in T, as measured by P, improves with experience E." In the scope of this book, we consider the supervised learning for classification applied, for example, to the problems where human experts exist, but where they are unable to explain their expertise. Specifically, in Part III we will demonstrate how to build classification solution in the cases of underwriters' decision-making process in insurance and financial industries.

There are several different types of machine learning algorithms (Dietterich, 2000). Supervised or semi-supervised learning, sometimes called inductive learning, builds models using training datasets that include both input data and desired outputs. When the problem relates to classification, the outputs are called labels. Using these outputs, supervised learning algorithms reveal a function that can predict the output associated with new inputs.

One of the examples that we will review in this book is when insurance underwriters exhibit expert-level abilities to evaluate an application for insurance and respond to the applicant with the most suitable offer, but they cannot accurately describe the steps they perform during the underwriting process and cannot share their knowledge and experience with young and less skillful underwriters. In this example, the historical data about underwriters' decision making can provide examples of the inputs and outputs so that machine learning algorithms can learn to model the underwriters' decisions. An example of application of supervised classification would be to develop a computer program that, when given an application for insurance in a specific insurance line of business, will determine what would be the applicant's response (accept or reject) to a specific insurance offer. Such a program is called a classifier, as it assigns a class (accept or reject) to an object (an applicant). The task of supervised learning is to construct a classifier given a set of already classified (labeled) examples like a set of insurance applications and offered insurance products, labeled with the acceptance or rejection responses. In the case of semi-supervised learning, desired outputs also contain missing values.

Unsupervised learning is the case when training data does not include desired outputs. An example of unsupervised learning is clustering where an algorithm attempts to find some structure in the data, like grouping or clustering of data points. Such algorithms learn from data that has not been labeled, classified, or categorized. In this case, it is hard to tell what good learning is and what is not.

The most ambitious type of learning is reinforcement learning. It is a type of machine learning technique that enables an algorithm to learn in an inter-active environment, by trial and error, using feedback from its own actions and experiences as a reward or punishment.

In this book, we discuss the optimization framework for only supervised machine learning methods.

2.2 Principles of Supervised Machine Learning

Sometimes, supervised learning is compared with what statisticians do. The "supervision" refers to the fact that the responses are available, and the term "learning" means any method or procedure that can be tested for the valid-ity of its output. When responses are available, the goals of analysis such as model identification, decision making, or prediction can be validated. More typically, supervised learning refers to classification.

The objective of classification is to assign objects of a population to one of several categories using a set of features observed on each object. For example, in the case of insurance underwriters, the set of features may include features of the insurance application and features of the insurance product that is offered in response to the application. The categories are known in advance and can be ordered or not ordered. In the same example related to insurance applications, the categories will be one of two: the offered insurance product will be accepted by the applicant, or the offer will be rejected.

Classifiers are built on training datasets, where both features and labels are known and available for the classifier-building algorithm.

The quality of a classifier is measured by calculating the ratios of correctly classified objects within each category. This measure is called the misclas-sification matrix and is calculated first on the training datasets and then on the testing datasets.

A misclassification matrix, sometimes, is also called a confusion matrix or an error matrix. It visualizes the performance of a machine learning method and serves the basis of calculating the metrics of classification performance. Each row of the misclassification matrix represents predicted labels, and each column represents actual labels. Each cell of this matrix contains the frequency of instances corresponding to the predicted and actual labels (see Table 2.1).

The metric of sensitivity, also called recall or hit rate, measures the true-positive rate (TPR):

$$\text{TPR} = \frac{\text{TP}}{P} = \frac{\text{TP}}{\text{TP} + \text{FN}} = 1 - \text{FNR} \qquad (2.1)$$

where
TP is the frequency of true-positive instances,
P is the number of actual positive cases,
FN is the frequency of false-negative instances,
FNR is the false-negative rate.

The metric of specificity, also called selectivity, measures the true-negative rate (TNR):

$$\text{TNR} = \frac{\text{TN}}{N} = \frac{\text{TN}}{\text{TN} + \text{FP}} = 1 - \text{FPR} \qquad (2.2)$$

where
TN is the frequency of true-negative instances,
N is the number of actual negative cases,
FP is the frequency of false-positive instances,
 FPR is the false-positive rate.

There are also other metrics that are calculated based on misclassification matrix, for example precision that measures positive predictive value (PPV), or false discovery rate (FDR), or other metrics.

The separate testing dataset contains both features and labels, but was not used for building the classifier. The reason we use the testing dataset is an attempt to address the key challenge of supervised learning: the problem of generalization. In fact, most machine learning methods build classifiers that

TABLE 2.1

Misclassification Matrix

Actual Predicted	Label "TRUE"	Label "FALSE"
Label "TRUE"	TP (true positive)	FP (false positive)
Label "FALSE"	FN (false negative	TN (true negative)
	P (condition positive)	N (condition negative)

produce very accurate classifications on the training datasets. However, we want to know the ability of the classifier to generalize for new data points (unseen data) that were not present in the training dataset.

The principles described above, however, leave out some important details. For example, each machine learning method has a set of hyperparameters, the values of which should be specified before building the classifier. These hyperparameters define the process of building a classifier, but their values cannot be inferred from training datasets. How can we specify the values of the hyperparameters that allow us to build the best classifiers?

Another important issue is the selection of the best classifier. Different classifiers can have very close ratios of correctly classified objects within each category. Let us say we need to build a classifier for the case with two output values, or labels, 0 and 1, and that the prediction of both values is equally important. Now assume that we have built two classifiers: the first one correctly predicting 90% of 0 and 85% of 1, while the second one correctly predicting 85% of 0 and 90% of 1. What criteria should be considered to choose the best classifier and do we, actually, need to choose only one?

Another important issue is the robustness of the classifier. Building and testing a classifier on available data does not guarantee the quality of the classifier applied to unseen data. What can be done to estimate the ability of a classifier to deal with previously unobserved data?

All these questions arise when we try to solve real-life problems using machine learning methods. Prior to attempting to answer them, let us review some of the machine learning methods.

2.3 Neural Network

2.3.1 Introduction

The neural network method was created as an attempt to mimic the way the human brain processes and stores information (Haykin, 1999). The basic component of the brain, the neuron, was discovered in 1836. It took more than 100 years till the modern era of neural network research and development was initiated in the classic paper "A Logical Calculus of the Ideas Immanent in Nervous Activity" of W.S. McCulloch and W. Pitts.

In the paper written in 1943, McCulloch, a neurophysiologist and cybernetician, and Pitts, a mathematician, proposed the first mathematical model of a neural network. The unit of this model, called a McCulloch–Pitts neuron, is still used in the field of neural networks.

A neural network creates connections between mathematical processing elements, neurons. Knowledge is encoded into the network through the strength (weights) of the connections between different neurons and by creating layers of neurons that work in parallel.

A neural network learns through the process of determining the number of neurons, or nodes, and adjusting the weights for the connections. The learning process uses a training dataset composed of input–output pairs and tries to find a function that produces the outputs based on the given inputs. The neural network approximates a function for the input domain derived from the training data.

There are two main types of neural networks: fixed (non-adaptive) and dynamic (adaptive). Fixed neural networks become set after completing the learning process. All structure components such as weights, connections, and node configurations become fixed, and the network turns into a function that does not change during the operation. The appropriate use of a fixed neural network would be a classification system in the conditions where the definition of inputs and outputs does not change, and the network is expected to perform the same classification repeatedly.

Dynamic neural networks never stop learning and continue to develop throughout their operation. They continue adapting to the most recent data and change their internal structure of neurons and weights. Dynamic neural networks are used in situations where a system needs to learn new data during its operation. This type of neural network is indispensable where we expect new input–output patterns, for example when the domains change over time, or outputs are impacted by slow-moving processes.

2.3.2 Neural Network Components

2.3.2.1 Activation Function

An important component of a neural network classifier is an activation function, which can be interpreted as an indicator function. The activation function of a neuron defines the output of that neuron depending on inputs to the neuron. Sigmoid functions are often used as the activation function. One example of a sigmoid function is a logistic function:

$$g(x) = \frac{1}{1 + \exp(-x)} \tag{2.3}$$

For this activation function, $x \geq 0$ activates the neuron.

Another example of the activation function used in neural networks is the symmetric sigmoid:

$$g(x) = 2 \times \frac{1}{1 + \exp(-x)} - 1 = \frac{1 - \exp(-x)}{1 - \exp(-x)} = \tanh\left(\frac{x}{2}\right) \tag{2.4}$$

More examples include the rectified linear unit (ReLU) (equation 2.5) and the radial basis function (equation 2.6):

$$g(x) = \begin{cases} 0, x \leq 0 \\ x, x > 0 \end{cases} \tag{2.5}$$

$$g(x) = \exp(-x^2) \tag{2.6}$$

For the radial basis activation function, $x = 0$ activates the neuron (Figure 2.1).

2.3.2.2 Neurons

The fundamental component of a neural network is a neuron. It is composed of an activation function g and a parameter vector $\vec{w} = (w_0, w_1, \ldots, w_n)$. A neuron takes a vector of real-value inputs, also called features, and calculates the

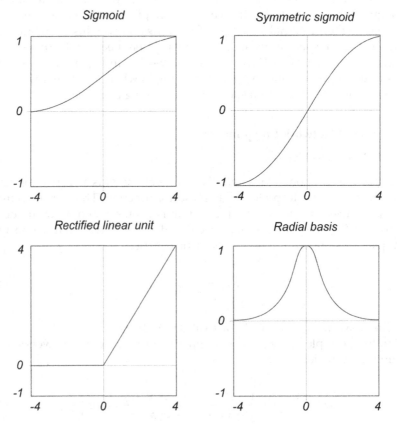

FIGURE 2.1
Activation functions.

output using the activation function and the parameter vector: if the result of the activation function is greater than the threshold (or equal to a specified value), then the output is 1; otherwise, the output is 0. More precisely, the output $o(x_1, \ldots, x_n)$ is computed by the neuron formulated by equation (2.7).

$$o(x_1, \ldots, x_n) = \begin{cases} 1 \text{ if } g(w_0 + w_1 x_1 + \cdots + w_n x_n) > 0 \\ 0 \text{ otherwise} \end{cases} \tag{2.7}$$

where

w_i is the weight that determines the contribution of input x_i to the neuron output

w_0 is the threshold that the weighted combination of inputs $(w_1 x_1 + \cdots + w_n x_n)$ must exceed to activate the neuron.

The learning of neuron involves estimating the values of the weights w_0, $w_1, \ldots, w_n$.

2.3.2.3 Networks

A neural network composes multiple neurons into a single function, and there are many different algorithms that can perform this job. Some of the neurons in a neural network use the vector of feature values as input, others use the output from some neurons as input, and yet others use a combination of the two. The parameters of a neural network include the parameters of the neurons g and $\bar{w}$, the way in which the neurons are connected, and the parameters of an output layer of linear functions. Neurons are typically arranged in layers according to what kind of input they receive. At the input layer, neurons receive feature vectors as input, and in subsequent layers, neurons receive outputs from previous layers as inputs. The layers other than the input and output are called hidden layers of neurons.

An example of a neural network is shown in Figure 2.2. This neural network has an output layer of m linear functions, $F_1, \ldots, F_m$, and a single hidden layer of k neurons, $Z_1, \ldots, Z_k$. Each neuron in the hidden layer receives

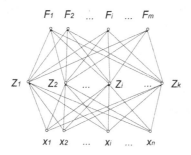

FIGURE 2.2
Neural network.

as input the feature vector $\vec{x} = (x_1, \ldots, x_n)$. The linear functions in the output layer receive as input the vector of values of the hidden neurons, $\vec{Z} = (Z_1, \ldots, Z_k)$.

2.3.3 R for Neural Networks

The package "neuralnet" developed by the team of Stefan Fritsch, Frauke Guenther, Marvin N. Wright, Marc Suling, and Sebastian M. Mueller contains the function neuralnet that implements several algorithms of training of neural networks such as backpropagation (Riedmiller, 1994), resilient backpropagation with or without weight backtracking, or the modified globally convergent version by Anastasiadis et al. (2005) and Taylor (2006).

This function is very popular among researchers; however, the usage of this function is not always correct. The biggest challenge of using this function is to assign values to its arguments. Table 2.2 describes the 19 arguments of the neuralnet function.

There are no recipes on how to choose values for the attributes required to train the neural network. For the argument "formula," the decision about what features to include in the model should be made. This problem by itself presents a significant challenge and is addressed in an extensive number of researches and publications. But whatever the approach the reader uses to choose features, this choice is confounded with the rest of the arguments of the function. For example, how to define the value of the argument "hidden": what is the needed number of hidden layers and the number of neurons in each layer? We can be sure that for a different set of features the answer will be different.

The choice of the one out of four available in the neuralnet function algorithms to be used for the neural network training is also difficult. Although we know how the algorithms work, the characteristics of the data we use to train the neural network are unknown in most cases. But then, even if we make some decision about the algorithm, how to choose the values for the attributes reflecting learning rate (learningrate.limit, learningrate.factor, learningrate)?

And if this is not enough, we must now decide about the forms of the loss function (err.fct) and the activation function (act.fct).

2.4 Support Vector Machine

2.4.1 Introduction

Support vector machine (SVM) is a proven classification method that directly minimizes the classification error without requiring a statistical

TABLE 2.2

Arguments of the "neuralnet" Function

	Argument	Description
1	formula	a symbolic description of the model to be fitted
2	data	a data frame containing the variables specified in the formula
3	hidden	a vector of integers specifying the number of hidden neurons in each layer
4	threshold	a numeric value specifying the threshold for the partial derivatives of the error function as stopping criteria
5	stepmax	the maximum steps for the training of the neural network
6	rep	the number of repetitions for the neural network's training
7	startweights	a vector containing starting values for the weights
8	learningrate.limit	a vector or a list containing the lowest and highest limit for the learning rate (for RPROP and GRPROP)
9	learningrate.factor	a vector or a list containing the multiplication factors for the upper and lower learning rate (for RPROP and GRPROP)
10	learningrate	a numeric value specifying the learning rate (for traditional backpropagation)
11	lifesign	a string specifying how much the function will print during the calculation of the neural network ("none," "minimal," or "full")
12	lifesign.step	an integer specifying the step size to print the minimal threshold in "full" lifesign mode
13	algorithm	a string containing the algorithm type to calculate the neural network
14	err.fct	a differentiable function that is used for the calculation of the error, or "sse" for the sum of squared errors, or "ce" for the cross-entropy
15	act.fct	a differentiable function that is used for smoothing the result of the cross-product of the covariate or neurons and the weights, or "logistic" for the logistic function, or "tanh" for a hyperbolic tangent
16	linear.output	If act.fct should not be applied to the output neurons, set linear output to TRUE, otherwise to FALSE
17	exclude	a vector or a matrix specifying the weights that are excluded from the calculation
18	constant.weights	a vector specifying the values of the weights that are excluded from the training process and treated as fix
19	likelihood	If the error function is equal to the negative log-likelihood function, the information criteria AIC and BIC will be calculated.

model. SVM can handle the massive feature space, and it is guaranteed to find a globally optimal separating hyperplane if it exists. Also, SVM classifiers are built on the boundary cases, which allow them to handle missing data. An SVM classifier can separate data that is not easily separable in the original space (for example, two-dimensional space) by mapping data into a higher-dimensional (transformed) space. Simply saying, a nonlinear classifier maps our data from the input space X to a feature space F using a nonlinear function $\Phi : X \rightarrow F$. In the space F, the discriminant function is as follows:

$$f(x) = w^T \Phi(x) + b \qquad (2.8)$$

2.4.2 Kernel

Kernel methods resolve the issue of mapping the data to a high-dimensional feature space. Suppose the weight vector can be expressed as a linear combination of the training examples:

$$w = \sum_{i=1}^{n} \alpha_i x_i \qquad (2.9)$$

Then

$$f(x) = \sum_{i=1}^{n} \alpha_i x_i^T x + b \qquad (2.10)$$

In the feature space F, this expression takes the form:

$$f(x) = \sum_{i=1}^{n} \alpha_i \Phi(x_i)^T \Phi(x) + b \qquad (2.11)$$

The feature space F may be high-dimensional, making this transformation difficult, unless the kernel function is defined in the form that can be computed efficiently:

$$K(x, x') = \Phi(x)^T \Phi(x') \qquad (2.12)$$

In terms of the kernel function, the discriminant function can be written as:

$$f(x) = \sum_{i=1}^{n} \alpha_i K(x, x') + b \qquad (2.13)$$

There are different kernel functions. For example, the polynomial kernel of degree d is defined as:

$$K(x, x') = \left(x^T x' + 1 \right)^d \qquad (2.14)$$

The feature space for this kernel consists of all monomials up to degree d, for example:

$$x_1^{d_1} x_2^{d_2} \ldots x_m^{d_m}, \text{where} \sum_{i=1}^{m} d_i \leq d \tag{2.15}$$

The increasing degree of the polynomial d leads to the increased flexibility of the classifier.

The other widely used kernel is the Gaussian (radial basis function) kernel defined by:

$$K(x, x') = \exp\left(-\gamma \|x - x'\|^2\right) \tag{2.16}$$

where $\gamma > 0$ is a parameter that controls the width of Gaussian. It plays a similar role as the degree of the polynomial kernel in controlling the flexibility of the resulting classifier.

Hyperbolic tangent, also known as the sigmoid kernel, is yet another popular kernel function. It is defined by the following formula:

$$K(x, x') = \tanh\left(\alpha x^T x' + b\right) \tag{2.17}$$

where α is sometimes referred to as the slope parameter and b as an intercept. It plays a similar role as the degree of the polynomial kernel in controlling the flexibility of the resulting classifier. It is interesting to observe that an SVM model with a sigmoid kernel function is equivalent to a two-layer perceptron neural network.

2.4.3 Margin

The margin is the length of the normal vector to the hyperplane from the two closest cases called support vectors; in other words, it is the margin with which the two classes are separated.

For a given hyperplane, we denote by $x_+ (x_-)$ the closest point to the hyperplane among the positive and negative cases. The norm (length) of a vector w is denoted by $\|w\| = \sqrt{w^T w}$. From a geometric perspective, the margin of a hyperplane f with respect to a dataset D can be defined as:

$$m_D(f) = \frac{1}{2} \hat{w}^T (x_+ - x_-) \frac{1}{\|w\|} \tag{2.18}$$

where $\hat{w} = w/\|w\|$ is the unit vector in the direction of w, assuming that x_+ and x_- are equidistant from the decision boundary.

2.4.4 Optimization

SVM uses a kernel function to find a hyperplane that maximizes the distance (margin) between two classes while minimizing training error. The maximum margin classifier is the discriminant function that maximizes the geometric margin $\frac{1}{\|w\|}$ which is equivalent to minimizing $\|w\|^2$. The soft-margin optimization problem (Cortes and Vapnik, 1995) is formulated as follows:

$$\text{minimize on } w,b : \frac{1}{2}\|w\|^2 + C\sum_{i=1}^{n} e_i \tag{2.19}$$

$$\text{subject to} : y_i\left(w^T x_i + b\right) \geq 1 - e_i, e_i \geq 0, i = 1,\ldots,n$$

where:

w is the margin,

C is the constant that sets the relative importance of maximizing the margin and minimizing the amount of slack,

e_i is the slack variables that allow a case to be in the margin ($0 \leq e_i \leq 1$, also called a margin error) or to be misclassified ($e_i > 1$), y_i – the output results of the training dataset, also called labels,

b is the bias that translates the hyperplane away from the origin.

2.4.5 Bias–Variance Trade-off and SVM Hyperparameters

The bias is an error caused by wrong assumptions in the learning algorithm. High bias means an underfitting between features and outputs. The variance is an error produced from variations in the sensitivity metric in response to small fluctuations in the training sets. High variance means overfitting between features and outputs.

The bias–variance trade-off is a central problem in supervised learning. In statistics and machine learning, the bias–variance trade-off is the property of a set of predictive models, where models with lower bias in parameter estimation have a higher variance of the parameter estimates across samples and vice versa. The bias–variance problem is the conflict in trying to simultaneously minimize these two sources of error that prevent supervised learning algorithms like SVM from generalizing beyond their training sets.

The quality of the results produced by SVM depends on the following hyperparameters:

- The type of kernel (radial, polynomial, sigmoid, etc.).
- The cost of misclassification (penalty) that is denoted as C in equation (2.19),

- The parameters of the kernel, for example γ in the cases of radial or sigmoid kernel transformations (equation 2.16) or d in the case of the polynomial kernel (equations 2.14 and 2.15).

A large C leads to low bias and high variance, while a small C leads to high bias and low variance. A large d or γ leads to high bias and small variance, while a small d or γ leads to low bias and high variance. For example, if γ is large (variance is small), it means that the SVM results do not have a wide influence and cannot be generalized beyond its training datasets.

2.4.6 R for SVM

The R package "e1071" was developed by a team of the Statistical Department of Vienna University of Technology. This package contains, among others, SVM function having 22 arguments.

TABLE 2.3

Arguments of the "svm" Function

	Argument	Description
1	formula	a symbolic description of the model to be fit
2	data	an optional data frame containing the variables in the model
3	x	a data matrix, a vector, or a sparse matrix
4	y	a response vector with one label for each row/component of x
5	scale	A logical vector indicating the variables to be scaled
6	type	svm can be used as a classification machine, as a regression machine, or for novelty detection
7	kernel	the kernel used in training and predicting
8	degree	the parameter needed for kernel of type polynomial
9	gamma	the parameter needed for all kernels except linear
10	coef0	the parameter needed for kernels of type polynomial and sigmoid
11	cost	the cost of constraints violation
12	nu	the parameter needed for nu-classification, nu-regression, and one-classification
13	class.weights	a named vector of weights for the different classes, used for asymmetric class sizes
14	cachesize	cache memory in MB (default 40)
15	tolerance	tolerance of termination criterion
16	epsilon	epsilon in the insensitive loss function
17	shrinking	option whether to use the shrinking heuristics
18	cross	a k-fold cross-validation on the training data
19	fitted	logical indicating whether the fitted values should be computed
20	probability	logical indicating whether the model should allow for probability predictions
21	subset	An index vector specifying the cases to be used in the training sample
22	na.action	A function to specify the action to be taken if NAs are found

Among these attributes, those of main interest for us relate to the specification of kernels and their hyperparameters: "kernel," "degree," "gamma," "coef0," and "cost." SVM function allows four kernels: linear, polynomial, Gaussian (radial basis function), and hyperbolic tangent (sigmoid). Depending on the choice of the kernel, the values of "degree," "gamma," and "coef0" should be specified. Unfortunately, the selection of a kernel and the values of hyperparameters very often is left as default, that is "kernel" = "radial," "degree" = 3, "gamma" = 1/data dimension, "coef0" = 0, and "cost" = 1. We wish the choice would be that easy!

2.5 SVM Modification Using Firth's Regression

2.5.1 Introduction

Because SVM does not model the data distribution, but instead directly minimizes the classification error, the output produced by SVM is a binary decision. In other words, SVM produces uncalibrated values that are not probabilities.

Consider the problem of binary classification: for inputs x, we want to determine whether they belong to one of two classes y, labeled +1 and –1. We assume that the classification problem will be solved by a real-valued function $f(x)$, by predicting a class label $y = \text{sign}(f(x))$. For many problems, it is convenient and often even necessary to get a probability $P(y = +1|f(x))$. This means that the classification gives not only an answer, but also a degree of certainty about the answer.

Although a binary decision is sufficient for many problems, it is difficult to arrive at a meaningful bias–variance trade-off. In addition, the application of an SVM classifier to new incoming data should produce a posterior probability $P(y = +1|f(x))$, i.e., the output of a classifier should be a calibrated posterior probability.

These concerns can be alleviated by mapping the output of SVM to probabilities. There are multiple approaches developed to address this problem. Vapnik (1998) suggested a method for mapping the output of SVM to probabilities by decomposing the feature space into a direction orthogonal to the separating hyperplane, and all of the N-1 other dimensions of the feature space. Platt (1999) proposed transforming the outputs of a classification model into a probability distribution over classes by fitting a logistic regression model to a classifier's output (see also Lin et al., 2007 and Murphy, 2012).

In this book, we propose applying Firth's penalized logistic regression to SVM output (Heinze, 2001). In other words, we first train an SVM and then estimate Firth's penalized logistic regression to map the SVM outputs into probabilities. The benefit of using penalized logistic regression to SVM output is the results are made more generalizable (for example, decrease overfitting). In classification, SVM along with penalized logistic regression creates output probabilities of class membership rather than point estimates produced by SVM alone. Rephrasing a classification problem as a probabilistic classification creates conditions for the application of bias–variance decomposition that was originally formulated for least-squares regression.

2.5.2 Logistic Regression

Let us consider logistic regression for a binary dependent variable y_i and vector of k covariates x_i with the corresponding vector of coefficients:

$$P\left(y_i = 1 \mid x_i, \beta\right) \equiv \pi_i = \frac{1}{1 + \exp\left(-x_i\beta\right)} \qquad (2.20)$$

where:
y_i is the binary dependent variable, $\vec{y} = (y_1, \ldots, y_n)$,
x_i is the vector of independent variables, $\vec{x_i} = (x_1, \ldots, x_k)$,
β is the vector of coefficients, $\vec{\beta} = (\beta_1, \ldots, \beta_k)$.
The log-likelihood for this model is the following:

$$\ln L\left(\beta \mid y\right) = \sum_{i=1}^{n}\left\{y_i \ln\left[\frac{1}{1 + \exp\left(-x_i\beta\right)}\right] + (1 - y_i)\ln\left[1 - \frac{1}{1 + \exp\left(-x_i\beta\right)}\right]\right\} \qquad (2.21)$$

Thus, the score function looks like:

$$\frac{\partial \ln L\left(\beta \mid y\right)}{\partial \beta} \equiv U(\beta) \sum_{i=1}^{n}\left\{y_i \ln\left[\frac{1}{1 + \exp\left(-x_i\beta\right)}\right] + (1 - y_i)\ln\left[1 - \frac{1}{1 + \exp\left(-x_i\beta\right)}\right]\right\} \qquad (2.22)$$

Using standard iterative methods (e.g., Newton–Raphson), we can find $\hat{\beta}$ by solving equation (2.20) for β.

2.5.3 Problem of Separation

A common problem in models for dichotomous dependent variables is "separation." Albert and Anderson (1984) coined this term, differentiating between "complete" and "quasi-complete" separation. Complete separation occurs when one or more of a model's covariates perfectly predict some binary outcome. Quasi-complete separation denotes the case in which such perfect prediction occurs only for a subset of observations in the data.

Separation means the existence of a sub-vector x_s by which all n observations can be correctly categorized as either $y_i = 0$ or $y_i = 1$. The existence of such a sub-vector leads to a monotonic log-likelihood. As a result, the maximum likelihood estimates $\widehat{\beta_s}$ for the variables in the x_s sub-vector equal positive or negative infinity, and the associated standard error estimates are infinite as well.

This is exactly the situation that happens in the SVM classifier outcome, where the value of $f(x)$ represents the distance to the separating hyperplane, and thus, its value perfectly predicts the binary outcome of classification.

There is extensive literature on the subject of separation, and multiple proposals on how to deal with this problem (See Rainey, 2016 and Zorn, 2005). In this book we will use Firth's bias-correction method to address the separation issue in logistic regression. Firth method that uses a penalized likelihood estimation is considered as an ideal solution for separation issue.

In his paper from 1993, Firth suggested a method for eliminating the small sample bias in maximum likelihood estimation. The main idea of Firth's approach is to introduce a bias term in the form of the Jeffreys' (1946) invariant prior into the standard likelihood function. As $N \to \infty$, the bias itself approaches zero, but for small N, the $O(N^{-1})$ bias is present. As a result, Firth created a penalized log-likelihood formula:

$$\ln L\left(\beta \mid y\right)^* = \ln L\left(\beta \mid y\right) + 0.5\ln\left|i(\beta)\right| \tag{2.23}$$

and the corresponding equation for score function:

$$\frac{\partial \ln L\left(\beta \mid y\right)^*}{\partial \beta} \equiv U\left(\beta\right)^* = U\left(\beta\right) + 0.5\text{tr}\left\{i(\beta)^{-1}\left[\frac{\partial i(\beta)}{\partial \beta}\right]\right\} \tag{2.24}$$

where $i(\beta)$ is Jeffreys' non-informative prior and $\ln L\left(\beta \mid y\right)$ and $U(\beta)$ are described by equations (2.21) and (2.22), respectively.

Firth also demonstrated that his penalized likelihood both is asymptotically consistent and eliminates the usual small sample bias for a broad class of generalized linear models. He also showed that penalized likelihood estimates exist in numerous situations in which standard likelihood-based estimates do not, including that of complete separation in binary response models. At the same time, it is obvious that as $N \to \infty$, Firth's penalized likelihood estimates converge to the MLEs under the usual regularity conditions.

There are studies that strongly suggest that the penalized likelihood approach to estimating logit models in the presence of separation is uniformly superior to its alternatives. For example, in their study, Galindo-Garre et al. (2005) applied Monte Carlo simulations to demonstrate that Firth's approach with Jeffreys' prior is superior both to the ad hoc correction of Clogg et al. (1991) and to Bayesian approaches that adopt other uninformative conjugate or normal priors.

While using penalized likelihood estimates, however, it is important to remember that the resulting penalized profile likelihoods for the coefficients are often asymmetrical. This means that the usual inferences based on Wald-type statistics can be misleading. Heinze and Schemper (2002) recommend the use of penalized likelihood ratio tests instead of standard Wald tests, along with a visual examination of each parameter's profile likelihood.

2.5.4 R for Firth's Regression

The first R package for Firth's penalized logistic regression, "brlr" package, was written by the author of the method, but currently, the "brlr" package is replaced by the more efficient and more comprehensive "brglm" package developed by Ioannis Kosmidis. This package contains functions that allow to fit binomial response GLMs using Firth's (1993) bias reduction method for the removal of the leading $O(N^{-1})$ term from the asymptotic expansion of the bias of the maximum likelihood estimator. The algorithm of fitting is described in Kosmidis (2007). The main function of the package, "brglm," has 20 arguments, which are described in Table 2.4.

In our case, Firth's penalized logistic regression will be applied to the outcome of SVM. The distance to the hyperplane calculated by SVM will be an independent variable and will be assigned to the argument "x," and the labels will be assigned to the argument "y."

Two other arguments we would like to mention here are "method" and "pl" as they allow specifying model estimation using Firth's penalized logistic regression approach. The default value of the "method" argument is "brglm.fit," which means either the modified-scores approach to estimation or maximum penalized likelihood will be used. The "pl" argument is assigned a logical value that indicates whether the model should be fitted using Firth's penalized likelihood with Jeffreys' invariant prior, or using the bias-reducing modified scores. The "pl" argument's default value is "FALSE" which results in estimates that are free of any $O(N^{-1})$ terms in the asymptotic expansion of their bias. When "pl" is assigned "TRUE" value, bias reduction is also achieved but, in general, not at such order of magnitude. In the case of logistic regression, the value of pl is irrelevant since maximum penalized likelihood and the modified-scores approach match for natural exponential families (Firth, 1993).

2.5.5 SAS for Firth's Regression

The LOGISTIC procedure of SAS software fits linear logistic regression models for discrete response data utilizing maximum likelihood method. It can also perform conditional logistic regression for data with binary response and exact logistic regression for data with binary and nominal response. The procedure executes maximum likelihood estimation with either the Fisher scoring algorithm or the Newton–Raphson algorithm. This procedure can

TABLE 2.4

Arguments of the "brglm" Function

	Argument	Description
1	formula	a symbolic description of the model to be fitted
2	family	a description of the error distribution and link function to be used in the model. "brglm" currently supports only the "binomial" family with links "logit," "probit," "cloglog," "cauchit"
3	data	an optional data frame containing the variables in the model
4	weights	an optional vector of "prior weights" to be used in the fitting process
5	subset	an optional vector specifying a subset of observations to be used in the fitting process
6	na.action	a function which indicates what should happen when the data contains NA
7	start	starting values for the parameters in the linear predictor
8	etastart	starting values for the linear predictor
9	mustart	starting values for the vector of means
10	offset	this can be used to specify an *a priori* known component to be included in the linear predictor during fitting
11	control.glm	control.glm replaces the control argument in "glm," but essentially does the same job. It is a list of parameters to control "glm.fit"
12	control	a list of parameters for controlling the fitting process
13	intercept	a logical value indicating whether an intercept should be included in the *null* model
14	model	a logical value indicating whether *model frame* should be included as a component of the returned value
15	method	the method to be used for fitting the model. The default method is "brglm.fit," which uses either the modified-scores approach to estimation or maximum penalized likelihood (see the pl argument below)
16	pl	a logical value indicating whether the model should be fitted using maximum penalized likelihood, where the penalization is done using Jeffreys' invariant prior, or using the bias-reducing modified scores
17	x	logical values indicating whether the response vector and model matrix used in the fitting process should be returned as components of the returned value
18	y	logical values indicating whether the response vector and model matrix used in the fitting process should be returned as components of the returned value
19	contrasts	an optional list, as in "glm"
20	control.brglm	a list of parameters for controlling the fitting process when method = "brglm.fit"

also perform the bias-reducing penalized likelihood optimization according to Firth (1993) and Heinze and Schemper (2002).

A complete description of PROC LOGISTIC can be found in SAS software documentation. Here we will discuss only the parameters related to Firth's method.

The MODEL statement of the PROC LOGISTIC specifies the model that should be estimated, and the list of options to specify how the model should be estimated. The model can be specified in one of two possible forms. The first form, referred to as "single-trial" syntax, is applicable to binary, ordinal, and nominal response data and is used when each observation in the dataset contains information about only a single trial, such as a single subject in an experiment.

The second form, referred to as "events/trials" syntax, is restricted to the case of binary response data. In this case, each observation contains information about multiple binary response trials in the form of the counts of observed subjects and the count of responding subjects.

In our case, the input data to PROC LOGISTIC will be in the "single-trial" form. An independent variable will be the distance value produced by SVM, and the dependent variable will be the label from the training or testing dataset.

To request that PROC LOGISTIC will perform Firth's penalized maximum likelihood estimation, the FIRTH option should be included in the MODEL statement.

2.6 Summary

In this chapter, we discussed the principles of machine learning and reviewed two specific methods: neural network and support vector machine. There is extensive literature reviewing these and other machine learning methods. However, there are multiple challenges to the practical application of these methods.

One such challenge is the selection of values of hyperparameters. The default values assigned, for example, in the above-discussed "svm" function written in R can be considered as place holders. Why "cost" is assigned 1? Or why "gamma" that is used for almost all kernels is $1/N$? Another challenge is the development of a robust classifier that would successfully deal with unseen data, that is the data not present either in training or in testing dataset.

In this book, we propose a framework that helps to address these challenges.

References

Albert, A., and Anderson, J. A. 1984. On the existence of maximum likelihood estimates in logistic regression models. *Biometrika*, 71 (1): 1–10.

Anastasiadis, A. D., Magoulas, G. D., and Vrahatis, M. N. 2005. New globally convergent training scheme based on the resilient propagation algorithm. *Neurocomputing*, 64: 253–270.

Clogg, C. C., Rubin, D. B., Schenker, N., Schultz, B., and Weidman, L. 1991. Multiple imputation of industry and occupation codes in census public–use samples using bayesian logistic regression. *Journal of the American Statistical Association*, 86: 68–78.

Cortes, C., and Vapnik, V. 1995. Support-Vector Networks. *Machine Learning*, 20: 273–297.

Dietterich, T. G. 2000. Machine learning. In *The Encyclopedia of Computer Science*, 4th Ed., eds. Hemmendinger, D., Ralston, A. and Reilly, E., 1056–1059. UK: John Wiley and Sons Ltd. Thomson Computer Press.

Firth, D. 1993. Bias reduction of maximum likelihood estimates. *Biometrika*, 80 (1): 27–38.

Galindo–Garre, F., Jeroen, K. V., and Wicher, P. B. 2004. Bayesian posterior estimation of logit parameters with small samples. *Sociological Methods and Research*, 33 (1): 88–117.

Haykin, S. 1999. *Neural Networks: A Comprehensive Foundation*. New York: Prentice Hall.

Heinze, G. 2001. The application of Firth's procedure to Cox and logistic regression. *Technical Report 10/1999 updated in January 2001*. Vienna: University of Vienna, Department of Medical Computer Sciences.

Heinze, G, and Schemper, M. 2002. A solution to the problem of separation in logistic regression. *Statistics in Medicine*, 21: 2409–2419.

Jeffreys, H. 1946. An invariant form for the prior probability in estimation problems. *Proceedings of the Royal Society A*, 186: 453–461.

Kosmidis, I. 2007. *Bias reduction in exponential family nonlinear models*. Ph.D. thesis. Department of Statistics, University of Warwick.

Lin, H., Lin, C., and Weng, R. C. 2007. A note on Platt's probabilistic outputs for support vector machines. *Machine Learning*, 68 (3): 267–276.

Mitchell, T. M. 1997. *Machine Learning*. Ohio, USA: McGraw Hill.

Murphy, K. P. 2012. *Machine Learning: A Probabilistic Perspective*. Cambridge: The MIT Press.

Platt, J. 1999. Probabilistic outputs for support vector machines and comparisons to regularized likelihood methods. *Advances in Large Margin Classifiers*, 10 (3): 61–74.

Rainey, C. 2016. Dealing with separation in logistic regression models. *Political Analysis*, 24: 339–355.

Riedmiller, M. 1994. Advanced supervised learning in multi-layer perceptrons – From backpropagation to adaptive learning algorithms. *Computer Standards & Interfaces*, 16 (3): 265–278.

Taylor, B. J. ed. 2006. *Methods and Procedures for the Verification and Validation of Artificial Neural Networks*. New York: Springer.

Vapnik, V. 1998. *Statistical Learning Theory*. New York: Wiley.

Zorn, C. 2005. A solution to separation in binary response models. *Political Analysis*, 13: 157–170.

3

Bootstrap Methods and Their Deployment in SAS and R

3.1 Introduction

The bootstrap is a resampling procedure used to estimate statistics on a population by sampling the original dataset. This method substitutes complicated and often inaccurate approximations of biases, variances, and other measures of uncertainty by computer simulations. In simple situations, the uncertainty of an estimate may be gauged by analytical calculations based on an assumed probability model for the available data. But in more complicated cases, this approach can be tedious and difficult, and its results are potentially misleading if inappropriate assumptions or simplifications have been made.

The idea of resampling procedures similar to the bootstrap has a long history, associated with the use of computers to perform a simulation. First, such procedures were mentioned in the early days of computing in the late 1940s. However, it was Efron (1979) who unified ideas and connected the simple nonparametric bootstrap (resampling the data with replacement) with statistical tools for estimating standard errors, such as the jackknife and delta methods. This first method is now commonly called the nonparametric IID bootstrap, the standard bootstrap that is appropriate for data that is either independent identically distributed (i.i.d.) or at least not serially dependent.

After the publication of the Efron (1982) monograph, the statistical and scientific community began to research these ideas actively and appreciate the extensions of the methods and their wide applicability. For a thorough overview of the bootstrap method and its theoretical background we refer the reader to Efron and Tibshirani (1993).

In our book, we show how to use the bootstrap approach in two different situations: first, for the purposes of minimizing bias in the input data, and second, for the purposes of generating multiple training and testing datasets. An overview of the bootstrap methods will help to explain the motivation of bootstrap application in the AI framework. The main idea of using multiple training and testing datasets is to achieve variability in the data used for building and validating classifiers. In this chapter, we also discuss what criteria should be used to stop generating of training and testing datasets.

3.2 Overview of Bootstrap Methods

The formal definition of Efron's bootstrap is as follows:

> Given a sample of n independent identically distributed random vectors $x_1, x_2, ..., x_n$ and a real-valued estimator $\hat{\theta} = \varphi(x_1, x_2, ..., x_n)$ of the parameter θ, a procedure to assess the accuracy of $\hat{\theta}$ is defined in terms of the empirical distribution function Fn. This empirical distribution function assigns probability mass $1/n$ to each observed value of the random vectors x_i, $i = 1, ..., n$.

The empirical distribution function is the maximum likelihood estimator of the distribution for the observations when no parametric assumptions are made. The bootstrap distribution for $\hat{\theta} - \theta$ is the distribution created by generating $\hat{\theta}$ values by independent sampling with replacement from the empirical distribution F_n. The bootstrap estimate of the standard error of $\hat{\theta}$ is then the standard deviation of the bootstrap distribution for $\hat{\theta} - \theta$. Almost any parameter of the bootstrap distribution can be used as a "bootstrap" estimate of the corresponding population parameter, for example the skewness, the kurtosis, or the median of the bootstrap distribution.

The practical application of the bootstrap technique usually requires the generation of samples obtained by independent sampling with replacement from the empirical distribution. Then, from the bootstrap sampling, an approximation of the bootstrap estimate is obtained using Monte Carlo method. The framework is quite straightforward:

1. Generate a sample with replacement from the empirical distribution (a bootstrap sample).
2. Compute θ^* the value of $\hat{\theta}$ obtained by using the bootstrap sample in place of the original sample.
3. Repeat Steps 1 and 2 k times.

Suppose that we want to estimate parameter θ that depends on a random sample $X = (x_1, x_2, ..., x_n)$. For example, θ can be a log variance of the sample, or a Pearson correlation coefficient. And assume that we have the following estimator of θ (that does not depend on the order of the values x_i):

$$\hat{\theta}(x_1, x_2, ..., x_n) = \varphi(x_1, x_2, ..., x_n) \tag{3.1}$$

However, we do not know the probability distribution of $\varphi(X)$ given θ. This means that we cannot estimate the error in estimating θ by $\varphi(X)$. Thus, we cannot tell if we can conclude $\theta \neq 0$ from an observed $\varphi(X) \neq 0$, no matter how large the value of the estimator is.

Let us see how the bootstrap approach can help to create a confidence interval for θ and to test the hypothesis $H_0:\theta = \theta_0$ using only the observed data $x_1, x_2, \ldots, x_n$.

3.2.1 The Basic Bootstrap

Let us say we have a sample $X = (x_1, x_2, \ldots, x_n)$ of size n. A bootstrap sample of X is a sample

$$X^* = \left(x_1^*, x_2^*, \ldots, x_n^*\right) \tag{3.2}$$

where each value x_i^* is a random sample from $X = (x_1, x_2, \ldots, x_n)$ with replacement. In other words,

$$P\left(x_j^* = x^*\right)\frac{1}{n}, i = 1, \ldots, n \tag{3.3}$$

where x_j^* are independent choices for $j = 1, \ldots, n$. Specifically, repeated values are allowed, and it is possible that:

$$x_{j1}^* = x_{j2}^* = x_i$$

From this, it follows that, given that X^* is of size n, some values in X can be left out.

We repeat this process k times and form k independent resamples: $x_{(1)}^* = x_{(2)}^* = x_{(k)}^*$. The bootstrap-resampled values for the estimator (equation 3.1) are as follows:

$$\hat{\theta}_i^* = \hat{\theta}\left(x_{(i)}^*\right), i = 1, \ldots, k \tag{3.4}$$

where the values $\hat{\theta}_i^*$ are treated as independent random samples with mean θ. The number of resampled values $\hat{\theta}_i^*$ is not limited by the sample size n.

The motivation behind this procedure can be explained in the following way. Let $F(z) = P(x_i \leq z)$ be the distribution function of the sample $(x_1, x_2, \ldots, x_n)$. Given this sample, our best estimate of $F(z)$ is the sample distribution function $\hat{F}(z)$ where each sample observation is assigned a weight of $1/n$. Thus, according to equation (3.3), each bootstrap sample x_i^* is a random sample of size n from $\hat{F}(z)$. Consequently, $\hat{\theta}(z)$ is a function of $\hat{F}(z)$. In other words, the process of obtaining $\hat{\theta}(X)$ from X can be described as shown in Figure 3.1.

Assuming that our best estimate of $F(z)$ is $\hat{F}(z)$ and that the bootstrap samples X^* are derived from $\hat{F}(z)$, we can modify Figure 3.1 as shown in Figure 3.2.

Thus, we can make inferences about the distribution of $\hat{\theta}(X)$ given θ by studying the distribution of $\hat{\theta}(X^*)$ given X. Now, we can simulate the

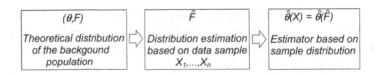

FIGURE 3.1
The process of obtaining sample estimator.

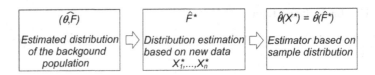

FIGURE 3.2
The process of obtaining bootstrap estimator.

distribution of $\hat{\theta}(X^*)$ by computing $\hat{\theta}_i^* = \left(X_{(i)}^*\right), i = 1,\ldots,k$, for k bootstrap samples $X_{(i)}^*$.

3.2.2 Hypothesis Tests, Estimates, and Confidence Intervals

Let us consider how the bootstrap approach helps in hypothesis testing. Theoretically, for any value of k, we can use the resampled values $\hat{\theta}_i^*$ and estimate a mean and variance for θ:

$$\hat{\theta}_{\text{mean}}^*(x) = 1/k \sum_{i=1}^{k} \hat{\theta}_i^* \tag{3.5}$$

$$\hat{V}_\theta(x) = \frac{1}{k-1} \sum_{i=1}^{k} \left\{ \hat{\theta}_i^* - \hat{\theta}_{\text{mean}}^*(X) \right\}^2 \tag{3.6}$$

and the corresponding 95% confidence interval for θ is:

$$\left(E_\theta(x) - 1.960 \sqrt{\frac{V_\theta(x)}{k}}, E_\theta(x) + 1.960 \sqrt{\frac{V_\theta(x)}{k}} \right) \tag{3.7}$$

Let us assume that the distribution of $\hat{\theta}(X)$ given θ is symmetric about θ for each θ. Then the distribution of $\hat{\theta}(X^*)$ will also be symmetric about θ, and a good estimator of θ will be

$$\hat{\theta}_{\text{med}}^*(x) = \text{median}\left\{ \hat{\theta}_i^*, i = 1,\ldots,k \right\} \tag{3.8}$$

Now, the 95% confidence interval for θ can be constructed as the upper and lower 2.5% quantiles of the sampled values $\hat{\theta}_i^*$. Thus, the bootstrap percentile 95% confidence interval for θ can be constructed as:

$$\left(\hat{\theta}_{(U)}^*, \hat{\theta}_{(k+1-U)}^*\right) \tag{3.9}$$

where

$\hat{\theta}_{(i)}^*$ are the values $\hat{\theta}_i^*$ sorted in ascending order so that $\hat{\theta}_{(1)}^* \le \hat{\theta}_{(2)}^* \le \cdots \le \hat{\theta}_{(k)}^*$.

U is equal to the $0.025k$ value, rounded down to the nearest integer (if $U = 0$, then it is replaced by 1).

The value 0.025 is calculated based on the given bootstrap percentile confidence interval, that is for 95% confidence interval this value is obtained as $(1 - 0.95)/2 = 0.025$.

For example, if $k = 1{,}000$, then $U = 25$ and $k + 1 - U = 976$, and the bootstrap percentile 95% confidence interval will look like $\left(\hat{\theta}_{(25)}^*, \hat{\theta}_{(976)}^*\right)$.

The advantages of equation (3.9) over equation (3.7) are as follows:

1. It does not require a theoretical (analytical) formula to calculate the confidence interval, which can be difficult to derive for complex estimators of θ.

2. The endpoints of the confidence interval (equation 3.9) are realizable values of $\theta = \hat{\theta}(X)$.

In addition, it is obvious that replacing θ by any monotonic function of θ replaces the bootstrap-sampled values for θ by the same monotonic function of those values. Thus, generating the median estimate (equation 3.8) and confidence interval (equation 3.9) for $\log(\theta)$ and then exponentiating the results leads to exactly the same results as if we had worked with θ in the first place. In general, if $g(\theta)$ is any monotonic function of θ, the following is true:

$$\widehat{g(\theta)}_i^* = g\left(\hat{\theta}_i^*\right) \tag{3.10}$$

where

$\widehat{g(\theta)}_i^*$ means the ith bootstrap-sampled value for $g\left(\hat{\theta}(X)\right)$.

This suggests yet the additional advantage of the bootstrap percentile confidence interval (equation 3.9) over the confidence interval based on normal distribution (equation 3.7). The proper scaling of θ is very often unclear or may seem to require transforming the values in equation (3.7) to eliminate outliers. Equation (3.10) means that we obtain the same median estimate (equation 3.8) and confidence intervals (equation 3.9) for any monotonic transformation of θ. It is important to keep in mind that the bootstrap percentile estimators (equations 3.8 and 3.9) were derived based on the assumption that the distribution of $\hat{\theta}(X)$ as a function of θ is symmetric about θ. The property (equation 3.10) weakens this assumption as now we can just assume the existence of a monotonic strictly increasing function $g(\theta)$ such that $g\big(\hat{\theta}(X)\big)$, as a function of θ is symmetrically distributed around $g(\theta)$, which creates a significantly large class of estimators.

3.2.3 Bias Reduction

Bootstrap allows us to estimate and then reduce bias.

The bootstrap estimator B^* of bias is $E\big(\hat{\theta}^* - \hat{\theta}\big)$, where $\hat{\theta}^*$ is an estimate of θ based on a bootstrap sample. An approximation of B^* is obtained by doing k bootstrap samples and then averaging the differences between the estimates $\hat{\theta}_i^*$ and the original sample estimate $\hat{\theta}$:

$$\hat{B}^* = \sum_{i=1}^{k}\big(\hat{\theta}_i^* - \hat{\theta}\big)/k \tag{3.11}$$

The purpose of estimating bias is to improve the biased estimator, for example by subtracting an estimate of its bias from it.

Suppose we observe that

$$\hat{\theta}_{\text{med}}^*(X) = \text{median}\big\{\hat{\theta}\big(X_{(i)}^*\big), \, i = 1, \ldots, k\big\} < \hat{\theta}(X) \tag{3.12}$$

Thus, we may state that using the estimate $\theta = \hat{\theta}_{\text{med}}^*(X)$ is not reasonable, as equation (3.12) may imply that the resampling process tends to yield smaller values of θ.

However, if the process of sampling X from the population is similar to the process of bootstrap sampling $X_{(i)}^*$ from X (which follows from Figure 3.2), then equation (3.12) should mean that $\theta > \hat{\theta}(X)$, rather than

$$\theta = \hat{\theta}_{\text{med}}^*(X) < \hat{\theta}(X)$$

One way of quantifying this median bias is to define "bias" as a proportion of bootstrap-sampled values $\hat{\theta}_i^*$ that are less than or equal to $\hat{\theta}(X)$:

$$\hat{B}_{\text{median}}^* = \frac{1}{k}\sum_{i=1}^{k}\mathbb{I}\big\{\hat{\theta}\big(X_{(i)}^*\big) \leq \hat{\theta}(X)\big\} \tag{3.13}$$

where

$\mathbb{I}$ is an indicator function that is equal to 1 when $\hat{\theta}\big(X_{(i)}^*\big) \leq \hat{\theta}(X)$, and 0 otherwise.

The expression in equation (3.12) can be interpreted as an approximation of $P\left(\hat{\theta}(X^*) \le \hat{\theta}(X)|X\right)$. If $\hat{B}^*_{median} = 1/2$, then there is no median bias about $\hat{\theta}(X)$ in the resampled values $X^*_{(i)}$. In some studies, threshoulds that indicate absence of median bias were suggested. For example,

$$0.40 \le \hat{B}^*_{median} \le 0.60 \text{ or } 0.35 \le \hat{B}^*_{median} \le 0.6$$

are suggested as rules indicating no concern of bias. Otherwise, it is suggested that a better-behaved estimator $\hat{\theta}(X)$ of θ should be found.

3.2.4 The Parametric Bootstrap

The original bootstrap was interpreted by Efron (1982) as a nonparametric maximum likelihood approach or, in other words, as a generalization of the maximum likelihood approach to the nonparametric framework. From this point of view, $\hat{F}(z)$ is a nonparametric estimate of $F(z)$. Thus, the ordinary nonparametric bootstrap estimates are "maximum likelihood" estimates.

If we assume that $F(z)$ has a parametric form such as, say, the Gaussian distribution, then the appropriate estimator for $F(z)$ would be a Gaussian distribution with the maximum likelihood estimates of μ and σ^2. Sampling with replacement from such a parametric estimate of $F(z)$ leads to bootstrap estimates that are maximum likelihood estimates in accordance with Fisher's theory. The Monte Carlo approximation to the parametric bootstrap is simply an approximation to the maximum likelihood estimate.

As shown in Figures 3.1 and 3.2, we approximated the unknown distribution (θ, F) by the sample distribution function $\hat{F}(z)$. Suppose we can assume that we know the distribution of X:

$$F(z) = F(\theta, z)$$

In this case, we substitute $\hat{F}(z)$ by $F\left(\hat{\theta}, z\right)$ in Figure 3.2. Thus, it means that in Figure 3.2 the bootstrap samples X^*_i are taken from $F\left(\hat{\theta}, z\right)$ instead of $\hat{F}(z)$.

Let us consider an example where we have a strong conviction that our data is distributed according to the following form of double exponential distribution:

$$f(X) = \frac{\exp\left(-|x - \mu|\right)}{2}$$

The maximum likelihood estimator of μ for this family is as follows:

$$\hat{\mu}(x_1, x_2, \ldots, x_n) = \hat{\mu}_m = \text{median}\{x_i, i = 1, \ldots, n\}$$

Following the process described in Figure 3.2, while substituting $\hat{F}(z)$ by $F(\hat{\theta}, z)$, the bootstrap samples $X^* = (x_1^*, x_2^*, ..., x_n^*)$ will now be independent random variables with density

$$f(X) = \frac{\exp\left(-\left|x - \hat{\mu}_m\right|\right)}{2}$$

Now, the bootstrap procedure uses more information about the distribution of X; thus, it should generate a more reasonable set of bootstrap-sampled values $\hat{\mu}_i^* = \hat{\mu}(X_{(i)}^*)$.

Using $F(\hat{\theta}, z)$ instead of $\hat{F}(z)$ in the bootstrap process presented in Figure 3.1 is called the parametric bootstrap, while using $\hat{F}(z)$ is called the nonparametric bootstrap.

3.2.5 m-out-of-n Bootstrap

As shown above, the basic nonparametric bootstrap extracts the bootstrap sample of the same size n as the size of the original sample. This is how it was first introduced by Efron. However, very soon the question of possibility to choose a bootstrap sample size m smaller than the size of the original sample n became a subject of research. Such a bootstrap has been called the "m-out-of-n" bootstrap and has been studied by various authors, for example Bickel et al. (1997).

The m-out-of-n bootstrap with m appropriately chosen is successfully used in cases where basic nonparametric bootstrap fails to be consistent. To produce consistent estimates, the asymptotic theory requires $m \rightarrow \infty$ as $n \rightarrow \infty$, but at a slower rate so that $m/n \rightarrow 0$. If we choose $m = o(n)$, the bootstrap mean is consistent. Athreya (1987) presented a proof of the convergence of the bootstrap mean when applying the m-out-of-n bootstrap. Amazingly, such a simple remedy works in a variety of examples, including dependent as well as independent observations. Such examples include the case when data has distribution with infinite moments, e.g., heavy tailed distribution, or the case of data distribution with extreme values, or yet another case of finite population, for instance survey sampling data.

3.2.6 Bootstrap Samples Similarity

While using the m-out-of-n bootstrap method to generate multiple training and testing datasets, we strive to increase the variability among sample datasets based on the randomness of observations selection. There is always a chance that two randomly selected samples are comprised from very similar subsets of observations, and the larger the sample size, the higher the chance of such similarity. Applying the same machine learning method to two very similar training datasets will most likely produce very similar classifiers.

By itself, it is not a problem, but our interest is to create different classifiers, and the estimation process is computationally expensive. We suggest applying the Jaccard similarity coefficient, also called the Jaccard index, to identify highly similar datasets and exclude or regenerate one of them.

Jaccard similarity coefficient is a statistic used to measure the similarity and diversity between finite-sample datasets. The Jaccard coefficient is defined as the size of the intersection divided by the size of the union of the sample sets:

$$J(A,\ B) = \frac{|A \cap B|}{|A \cup B|} = \frac{|A \cap B|}{A + B - |A \cap B|} \tag{3.14}$$

where

$|A \cap B|$ is the size of the intersection between A and B datasets,
$|A \cup B|$ is the size of the union of the A and B sample datasets.

Jaccard similarity coefficient statistic is limited: $0 \le J(A,\ B) \le 1$, and when A and B datasets are both empty, it is defined as $J(A,\ B) = 1$.

Another statistic, the Jaccard distance, measures the dissimilarity between two sample datasets. It is complementary to the Jaccard coefficient and is obtained by subtracting the Jaccard coefficient from 1, or, likewise, by dividing the difference of the sizes of the union and the intersection of the two sets by the size of the union:

$$D_J(A,\ B) = 1 - J(A,\ B) = \frac{|A \cup B| - |A \cap B|}{|A \cup B|} \tag{3.15}$$

Using $J(A,\ B)$ or $D_J(A,\ B)$ allows identifying pairs of similar bootstrap samples. Then one of these samples can be excluded from the further process of building a classifier.

3.3 Implementation of Bootstrap in SAS and R

3.3.1 m-out-of-n in SAS

In SAS, bootstrap can be implemented using the SURVEYSELECT procedure that includes a variety of methods for selecting probability-based random samples. Using this procedure, we can select a simple random sample or can sample according to a complex multistage sample design applying stratification, clustering, and unequal probabilities of selection.

The statements available in the PROC SURVEYSELECT are summarized in Table 3.1.

Using PROC SURVEYSELECT, we can deploy the m-out-of-n bootstrap method. For example, let us create 100 samples of 1,000 values with replacement from a "data_index" dataset that contains observation numbers of the input dataset:

```
proc surveyselect data=data_index
    method=urs n=1000 reps=100
    seed=123 out= data_index_sample;
run;
```

As a result of this procedure, the output "data_index_sample" dataset contains 100 samples of size 1,000. The method "urs" stands for "unrestricted random sampling" which means selection with equal probability and with replacement. In the case of more than a single selection of the same observation in the sample, the output dataset contains a single copy of this observation, but the variable "NumberHits" records the number of selections for each observation.

3.3.2 m-out-of-n in R

In R, the bootstrap can be implemented using two functions: "sample" and "replicate".

The "sample" function generates a sample of the specified size from the dataset or elements, either with or without replacement. The syntax is:

```
sample(x, size, replace = FALSE, prob = NULL)
```

TABLE 3.1

Parameters of SURVEYSELECT Procedure

PROC SURVEYSELECT options	invokes the procedure and optionally identifies input and output datasets; specifies the selection method, the sample size, and other sample design parameters
STRATA variables </ options>	The optional statement that identifies a variable or set of variables that stratify the input dataset
SAMPLINGUNIT | CLUSTER variables </ options>	The optional statement that identifies a variable or set of variables that group the input dataset observations into sampling units (clusters)
CONTROL variables	The optional statement that identifies variables for ordering units within strata
SIZE variable	identifies the variable that contains the size measures of the sampling units; it is used for probability proportional to size selection method
ID variables	

TABLE 3.2

Arguments of the "sample" Function

Argument	Description
x	dataset or a vector of one or more elements from which the sample is to be chosen
size	size of the sample
replace	"TRUE" or "FALSE" value depending on whether the sampling should be with or without replacement
prob	an optional vector of probability weights for obtaining the elements of the vector being sampled

The parameters of the function are described in Table 3.2.

The sample function is most commonly used to take a sample of the elements of a vector, which can be done either with or without replacement.

The "replicate" function repeats a given function a specified number of times. The syntax is the following:

```
replicate(n, expr)
```

where "n" is the number of desired replications and "expr" is the expression to be replicated.

Using these functions, the deployment of m-out-of-n bootstrap method is straightforward. As in the example used for PROC SURVEYSELECT, let us create 100 samples of 1,000 values with replacement from a "data_index" vector containing indexes of the input dataset:

```
set.seed(123)
MNBootstrap<-replicate(100, sample(data_index, 1000, replace =
TRUE))
```

In Part II of the book, we will demonstrate the usage of these functions in the AI framework.

3.4 Summary

In the machine learning process of building classifier, one of the largest challenges is to make the classifier ready for unseen data. In this chapter, we reviewed the bootstrap methods that allow us to create multiple training and testing datasets from the original labeled data. The main reason for creating these datasets is to allow variability of data for the process of classification, and as a result, to create the classifier (or classifiers) that successfully deal with such variability. We also use bootstrap methods at the stage of data cleansing to minimize bias in the input data.

References

Athreya, K. B. 1987. Bootstrap estimation of the mean in the infinite variance case. *The Annals of Statistics*, 15: 724–731.

Bickel, P. J., Gotze, F., and van Zwet, W. R. 1997. Resampling fewer than n observations: gains, losses, and remedies for losses. *Statistica Sinica*, 7: 1–31.

Efron, B. 1979. Bootstrap methods; another look at the jackknife. *The Annals of Statistics*, 7: 1–26.

Efron, B. 1982. The jackknife, the bootstrap, and other resampling plans. *CBMS-NSF Regional Conference Series in Applied Mathematics*. Vol. 38. Stanford, California: Stanford University.

Efron, B., and Tibshirani, R. J. 1993. *An Introduction to the Bootstrap*. New York: Chapman and Hall.

4

Outliers Detection and Its Deployment in SAS and R

4.1 Introduction

Machine learning process, regardless of the method used, depends heavily on the so-called input datasets. It is the most crucial aspect that makes algorithm training and testing possible. A dataset is a collection of data. In other words, a dataset corresponds to the contents of a single data table, or a single statistical data matrix $X(n \times p)$ where each column $j = 1,\ldots, p$ of the table represents a particular variable (feature) and each row $i = 1,\ldots, n$ corresponds to a specific observation of the dataset.

An input dataset serves as a source of both training and testing datasets. The training dataset is used to train an algorithm to apply machine learning concepts, such as neural networks and support vector machine, to learn and produce results. The training dataset includes the so-called labeled observations, which consist of both input data and the expected output (labels). The test dataset is used to evaluate how well the algorithm was trained with the training dataset. In machine learning projects, the goal of the testing stage is to evaluate how well the classifier, created in the process of training the algorithm, can work with previously unseen data. Thus, the training dataset cannot be used in the testing stage because the algorithm will already know in advance the expected output.

Because input datasets are usually flawed, data cleansing is a very important step in the machine learning process. To put it in a nutshell, data cleansing is a set of procedures that helps to make a dataset more suitable for machine learning. To put it simply, the quality of training data determines the performance of machine learning algorithms. Let's review the methods of data cleansing such as detection and treatment of outliers and bias correction.

4.2 Outliers Detection and Treatment

It is difficult to detect outliers in p-variate data when $p > 2$ because one can no longer rely on visual inspection of data. Although it is relatively easy to detect a single outlier using, for example, the Mahalanobis distances, this approach no longer suffices for multiple outliers, due to the masking effect when the outliers do not necessarily have large Mahalanobis distances. Donoho (1982) and Donoho and Gasko (1992) demonstrated that many methods for estimating multivariate location parameter break down in the presence of $n / (p + 1)$ outliers, where n is the number of observations in the dataset and p is the number of variables. We propose to use the minimum covariance determinant (MCD) method of Rousseeuw (1985) to detect multiple outliers in a multivariable setting.

4.2.1 Minimum Covariance Determinant Method

The objective of the MCD method is to find h out of n observations whose covariance matrix has the lowest determinant. The MCD estimate of mean is the average of these h observations, and the MCD estimate of variance is their covariance matrix. MCD has an important advantage. According to Butler et al. (1993), it is statistically efficient because the MCD is asymptotically normal. Regardless of this advantage, MCD has not been commonly applied for detecting outliers due to its computational complexity. The newly developed fast algorithm of MCD (Rousseeuw and Van Driessen, 1999) greatly outperforms other methods in terms of statistical efficiency and computational speed. Also, MCD can deal with a training dataset of a large size, that's why we recommend the MCD method to detect outliers in multivariable datasets. The motivation of the MCD algorithm can be explained as shown below.

Suppose we take a random sample $X_{(h)}$ of size h out of an input dataset $X = (x_1, x_2, \ldots, x_n)$. We can evaluate the similarity between data points in the full dataset and our randomly sampled subset. Let $M = (\mu_1, \mu_2, \ldots, \mu_p)$ be the mean of the random subset and S be the standard covariance matrix of the random subset. We use the Mahalanobis distance D to estimate similarity:

$$D = \left[(X - M)^T S^{-1} (X - M) \right]^{1/2} \tag{4.1}$$

The basic idea of the algorithm is to start with small subsets of size $p + 1$, their average M, covariance matrix S, and a corresponding h-subset of the h observations with the smallest Mahalanobis distances.

For a detailed description of the Fast-MCD algorithm, we refer the readers to the work of Rousseeuw and Van Driessen (1999). The newly developed algorithm, by Boudt et al. (2018), deals with the situation where the number of variables (features) exceeds the dataset size. So, when the size of the training dataset is smaller than the number of variables (features), the MCD method should be replaced by the minimum regularized covariance determinant (MRCD) method.

4.2.2 MCD in SAS

The Fast-MCD algorithm is implemented in different software packages such as R, SAS, and Python. However, the implementation of this algorithm can be different depending on what it is used for. The R-written function CovMcd from the robustbase package, for example, calculates scalable robust estimators using the MCD algorithm, but does not produce dataset cleaned from outliers.

The SAS software has the ROBUSTREG procedure that is usually used as a regression procedure. However, PROC ROBUSTREG can also compute MCD estimates. The statements available in PROC ROBUSTREG are summarized in Table 4.1.

In the context of a regression model, the word "outlier" means an observation where the value of the response variable is far from the predicted value. In contrast, PROC ROBUSTREG uses the MCD algorithm to identify multivariate outliers in the X space, i.e., outliers in the covariate space of explanatory variables. Such outliers are also referred to as leverage points. We can

TABLE 4.1

Parameters of the ROBUSTREG Procedure

PROC ROBUSTREG < options >	Invokes the Procedure
BY variables	Allow obtaining separate analyses of observations in groups defined by the BY variables.
CLASS variables	Specify which explanatory variables to treat as categorical.
EFFECT name=effect-type(variables < / options >)	Enables constructing special collections of columns for design matrices.
ID variables	Name variables to identify observations in the outlier diagnostics tables.
MODEL response=< effects >< / options >	Specifies the variables to be used in the regression.
OUTPUT<OUT=SAS-data-set>keyword=name<... keyword=name >	Creates an output dataset that contains final weights, predicted values, and residuals.
PERFORMANCE<options >	Tunes the performance of the procedure by using single or multiple processors available on the hardware.
TEST effects	Request robust linear tests for the model parameters.
WEIGHT variable	Identifies a variable in the input dataset whose values are used to weight the observations.

use PROC ROBUSTREG to create an output dataset where the observations identified as outliers by the MCD algorithm will be flagged. As the MCD estimates are produced for the explanatory variables, a response variable is not important. We can use the label variable or can simply generate a random response variable.

Suppose we have a dataset X that has 10 columns named X1, ..., X10 that represent independent variables, and a column Y that represents labels. To generate an output dataset with outliers identified by the MCD method, we will use the LEVERAGE(MCDINFO) option in the MODEL statement.

```
proc robustreg data=X method=lts;
   model y = x1 - x18 / leverage(MCDInfo);
   output out=x_output leverage=leverage;
run;
```

The "x_output" dataset will include all variables from the original X dataset, as well as the "leverage" indicator variable with values 0 and 1. When the "leverage" variable equals 1, it flags the observations that are far from the center of the explanatory variables. They are multivariate "outliers" in the space of the X variables.

Of note, PROC ROBUSTREG excludes observations with missing values from calculating estimations. If a different treatment of missing values is preferable, for example some type of imputation, then it should be done before applying PROC ROBUSTREG to the data.

4.3 Bias Reduction

The dataset from which outliers were removed using the above-described method serves as an input for further data cleansing process. The main goal of the next stage of the cleansing process is the identification of bias, if any, and elimination of observations that are responsible for that bias. We propose to use a bootstrap approach for bias determination (see 3.2.3 Bias Reduction). For each variable in the dataset, the approach works as follows:

1. Sample with replacement a subset from the dataset. The size of the subset can be the same as that of the original dataset, or smaller when using the m-out-of-n bootstrap method (see 3.3 m-out-of-n Bootstrap)

2. Compute an estimate of interest (sample mean $\hat{\mu}_1$ or median $\hat{m}_1$)

3. Perform steps (1) and (2) n times where n can be as large as 1,000. This will create n bootstrap estimates, e.g., $\hat{m}_1, ..., \hat{m}_n$

4. Compute the mean of the n bootstrap estimates

The difference between the estimate computed using the original dataset and the mean of the n bootstrap estimates is a bootstrap estimate of a bias $\hat{B}$:

$$\hat{B} = \frac{1}{n}\sum_{i=1}^{n}\hat{m}_i - \hat{m}$$

5. To correct the bias of the estimate, subtract the bootstrap bias estimate from the estimate of interest calculated from the training dataset. For example, if the estimate of interest is median, then the bias-corrected estimate will be:

$$\hat{m}^* = \hat{m} - \left(\frac{1}{n}\sum_{i=1}^{n}\hat{m}_i - \hat{m}\right)$$

which is the same as:

$$\hat{m}^* = 2\hat{m} - \frac{1}{n}\sum_{i=1}^{n}\hat{m}_i$$

If bias is present, the observations that are "responsible" for the bias should be removed from the dataset. It can be done by calculating the corresponding confidence interval based on the bias-corrected estimate $\hat{m}^*$ or $\hat{\mu}^*$. Observations whose values for a specific variable are outside the confidence interval are removed from the dataset. This process can be easily coded in R or SAS.

4.4 Summary

In this chapter, we discussed the importance of cleansing data for the use in the machine learning process. Specifically, we outlined two important steps that take care of multidimensional outliers and bias. Additional steps of cleansing may be required, which could be dictated by the nature of the data used for classification. This may include rule-based exclusion of erroneous values or treatment of missing values using different ways of imputation.

References

Boudt, K., Rousseeuw, P., Vanduffel, S., and Verdonck, T. 2018. The minimum regularized covariance determinant estimator. 10.2139/ssrn.2905259.

Butler, R. W., Davies, P. L., and Jhun, M. 1993. Asymptotics for the minimum covariance determinant estimator. *Annals of Statistics*, 21 (3): 1385–1400.

Donoho, D. L. 1982. *Breakdown Properties of Multivariate Location Estimators*. Ph.D. qualifying paper. Department of Statistics, Harvard University.

Donoho, D. L., and Gasko, M. 1992. Breakdown properties of location estimates based on halfspace depth and projected outlyingness. *The Annals of Statistics*, 20 (4): 1803–1827.

Rousseeuw, P. 1985. Multivariate estimation with high breakdown point. *Mathematical Statistics and Applications*, 8 (283–297): 37.

Rousseeuw, P. J., and Van Driessen, K. 1999. A fast algorithm for the minimum covariance determinant estimator. *Technometrics*, 41 (3): 212–223.

5

Design of Experiments and Its Deployment in SAS and R

5.1 Introduction

Machine learning process of classification depends on multiple conditions, and selection of these conditions to achieve the best result presents a significant challenge.

We start building a classifier when we have a dataset where each observation is assigned a label (output). The first question we usually encounter is what features to use for building a classifier. Feature selection is a difficult question that may require deep knowledge of the problem domain. We may choose different approaches to create reasonable feature sets (Dong and Liu, 2018, Weston et al, 2000), but as a result, we end up with multiple possibilities. In the process of building the classifier, we need to decide which of these feature sets works best for our classification problem.

As the next step, we need to make our mind about the machine learning method that we will use to build the classifier. Suppose we have reasons to choose a specific method. However, if this is not the case, then we have several methods we want to try for building the classifier.

As discussed in Chapter 2, using machine learning methods requires assigning values to multiple hyperparameters. For example, in the case of neural network, a decision should be made about an activation function, set of neurons, a number of hidden levels, etc. And in the case of support vector machine, we need to choose a kernel and its parameters, as well as the value of the cost parameter.

It would be unreasonable to assume that there is only one set of hyperparameters that permits to build the best classifier. As well, the definition of the "best" classifier can vary from case to case. As discussed in Chapter 2, the quality of a classifier is estimated based on the confusion (misclassification) matrix. There are different metrics that can be derived from this matrix, and the decision of what metrics should be used to evaluate the quality of a classifier and what value of the metrics identifies a good classifier should be made prior to building the classifier.

An additional measurement of a classifier quality is its robustness to data contamination. Data contamination helps heuristically evaluate the robustness of a classifier toward unseen data and find a solution of bias–variance trade-off.

Finding an answer to each of the above-stated questions is challenging in itself. But to make the situation more difficult, an answer to one question depends on answers to other questions. For example, different feature sets may produce better results with different kernels. To find a combination of conditions that deliver the best classification results, we will use the design of experiments (DoE).

5.2 Application of DoE in AI Framework

The AI framework described in this book utilizes the DoE methodology in order to find the best classifiers according to metrics defined for each case. Application of DoE includes defining factors and their levels, the structure of the experiment design, and the response variable measured on each experiment. Below, we will review the design of experiment (DoE) methodology.

A/B testing or "one factor at a time" (OFAT) was a very popular scientific method until the early twentieth century. In this method, one variable/factor is tested at a time while all other conditions of the experiment are considered not changing. However, at the beginning of the twentieth century, Sir Ronald Fisher introduced the concept of applying statistical analysis during the planning stages of research rather than at the end of experimentation. In 1935, his book "The Design of Experiments" established the foundation for the DoE as a statistical discipline. Cox (1958) presents a comprehensive nonmathematical interpretation of the design and analysis of experiments. Steinberg and Hunter (1984) state that Fisher's "monumental work was guided by the key insight that statistical analysis of data could be informative only if the data themselves were informative, and that informative data could best be assured by applying statistical ideas to the way in which the data were collected in the first place." This is exactly what we want to achieve: to assure that through building a classifier, we collect informative data to derive informative insights about the classifier quality.

The methodology of the design of experiments uses special terminology and is based on several main principles, which are described in vast existing literature like Cochran and Cox (1997), Dean and Voss (1999), Hicks and Turner (1999), Hinkelmann and Kempthorne (1994), Kempthorne (1952) and more. We briefly describe the terminology and the principles below.

5.2.1 Terminology of DoE

5.2.1.1 Experiment

An experiment is an action when treatment is deliberately imposed on a group of objects or subjects with the purpose of observing the response. This is in contrast to an observational study, when collecting and analyzing data does not involve a deliberate change of existing conditions.

5.2.1.2 Experimental Unit

An experimental unit is a physical entity on which treatment is imposed. The experimental unit is also the unit of statistical analysis.

5.2.1.3 Factor

A factor of an experiment is a controlled independent variable. The values (or levels) of such a variable are set by the experimenter. In the AI framework, a degree of polynomial kernel is one of the factors that can have values on three levels such as 2, 3, or 4. Another example of a factor is the cost variable that can be defined on three levels. For example, level 1 means that the value of the cost variable belongs to an interval [1; 5), level 2 means that the value of the cost variable belongs to an interval [5; 20), and level 3 is linked with an interval [20, 100).

5.2.1.4 Treatment

Treatment is something that researchers administer to experimental units. Treatment is defined as a combination of factor levels. In the AI framework, a treatment means conditions in which we build a classifier. In continuation of the example above, a polynomial kernel with the degree of four and with the cost variable in the interval [5; 20) defines the treatment, or, in other words, the conditions in which a classifier will be built.

5.2.2 Principles of DoE

5.2.2.1 Randomization

Randomization is the process of assigning treatments to experimental units, so that each unit has the same chance to receive treatment. The assignment of treatments to experimental units at random distinguishes a designed (planned) experiment from an observational study or so-called "quasi-experiment." There is an extensive body of statistical theory exploring the consequences of the allocation of units to treatments using some random number generator. Assigning units to treatments randomly tends to decrease confounding, which appears due to latent factors related to the treatment. It is only if the experimental units are a random sample from a population that the results of the experiment can be applied reliably from the experimental units to the larger statistical population The probable error of such an extrapolation depends on the sample size, among other things, and can be estimated.

5.2.2.2 Statistical Replication

Measurements are subject to measurement uncertainty, and thus have variation. The same measurements are repeated to improve accuracy—in other words, replicated. Replications help to identify the sources of variation, better estimating the true effects of treatments, improving the experiment's reliability and validity.

5.2.2.3 Blocking

In some cases, full randomization of treatments is not feasible, or a researcher may believe that experimental units should be grouped together to decrease variability within a group. Blocking is the non-random arrangement of experimental units into groups called blocks or lots. Blocking reduces variation between units, which is produced by known but irrelevant sources. Thus, blocking leads to greater precision in the estimation of the source of variation in the experiment. In a block design, experimental units are first divided into homogeneous blocks before they are randomly assigned to treatment groups.

5.2.2.4 Orthogonality

The concept of orthogonality is important in the design of experiments as it guarantees that the effects of different factors and interactions can be estimated in the model separately from each other. Thus, analysis of an orthogonal design is straightforward as each main effect and interaction can be estimated independently. The estimates for the effects and coefficients remain unchanged even if some factors or interactions are removed from the model. However, if a design is not orthogonal, either by plan or by accidental loss of data, the analysis must be carried out very carefully and the interpretation of the results is not as straightforward.

5.2.3 Full-Factorial Experiment

In statistics, a full-factorial experiment is defined as an experiment with a design consisting of two or more factors, each with two or more discrete levels, so that all possible combinations of the levels across all the factors are applied to the experimental units (Box et al., 1978). All the possible combinations of the levels across all the factors in a full-factorial design can be presented on a hypercube in the k-dimensional design space defined by the minimum and maximum levels of each of the k factors. The experimental points are sometimes called factorial points. The total number of experimental runs is calculated as a product of the numbers of levels of the factors. Let's say we conduct a full-factorial experiment with three factors A, B, and C, where A, B and C are measured on three levels. Then, the experiment has $3^3 = 27$ runs, which are shown as a hypercube in Figure 5.1.

The full-factorial 3^3 design of experiment is shown in Table 5.1.

If there are m replicates in an experiment, then each run will be replicated m times. Replications are required to estimate an error from executing the experiment under exactly the same conditions.

A decision about a number of levels to be used for each factor is based on assumptions about the relations of the factor and the response variable. For example, one of the assumptions for factors set at two levels is that the value of the response variable is approximately linear over the range of the factor settings. Two-level designs are the lowest level designs that meet the basic criteria of designs—efficiency with respect to the number of runs and the ability to identify significant effects. As mentioned above, designs with factors on two levels can only describe linear (straight line) relations. Designs with factors on three levels allow for a check of main effects linear and quadratic behavior, as well as for the examination of interaction terms involving squares of the factors of interest. In case there is a reasonable assumption that the relation between a factor and the response has a cubic nature, and this assumption needs to be investigated in the experiment, then the factor should be defined on four levels.

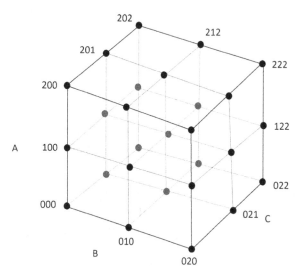

FIGURE 5.1
Hypercube for 3^3 full-factorial experiment.

TABLE 5.1

Full-Factorial 3^3 Design of Experiment

Run	A	B	C	Run	A	B	C	Run	A	B	C
1	0	0	0	10	1	0	0	19	2	0	0
2	0	0	1	11	1	0	1	20	2	0	1
3	0	0	2	12	1	0	2	21	2	0	2
4	0	1	0	13	1	1	0	22	2	1	0
5	0	1	1	14	1	1	1	23	2	1	1
6	0	1	2	15	1	1	2	24	2	1	2
7	0	2	0	16	1	2	0	25	2	2	0
8	0	2	1	17	1	2	1	26	2	2	1
9	0	2	2	18	1	2	2	27	2	2	2

Full-factorial experiments allow the investigator to study the effect of each factor on the response variable, as well as the effects of all interactions between factors on the response variable. The analysis of the experimental data can be performed using different methods, for example, multiple linear regression that models relationship between dependent variable (response) y and independent variables (factors) and their interactions, or ANOVA (analysis of variance) that explains how the total variation in data is split into variation explained by the regression model and the residual variation (Wu and Hamada, 2000, 2009).

Let's consider the example of 3^3 design of experiment presented in Table 5.1 with three factors A, B, and C measured on three levels. Suppose that the number of replications is m. So, the number of factor–level combinations in the experiment is equal to $3 \times 3 \times 3 = 27$. The linear model will look like the following:

$$y_{ijkl} = \mu + \alpha_i + \beta_j + \gamma_k + (\alpha\beta)_{ij} + (\alpha\gamma)_{ik} + (\beta\gamma)_{jk} + (\alpha\beta\gamma)_{ijk} + \varepsilon_{ijkl} \qquad (5.1)$$

where

$i, j, k = 1, 2, 3$ and $l = 1, \ldots, m$,

$\alpha_i, \beta_j,$ and γ_k are the A, B, and C main effects, respectively,

$(\alpha\beta)_{ij}, (\alpha\gamma)_{ik},$ and $(\beta\gamma)_{jk}$ are the A × B, A × C, and B × C two-factor interaction effects, respectively,

$(\alpha\beta\gamma)_{ijk}$ is the A × B × C three-factor interaction effect,

ε_{ijkl} are independent errors distributed as $N(0, \sigma^2)$.

The ANOVA derivation is based on the following zero-sum constraints:

$$\sum_{i=1}^{3}\alpha_i = \sum_{j=1}^{3}\beta_j = \sum_{k=1}^{3}\gamma_k = 0$$

$$\sum_{i=1}^{3}(\alpha\beta)_{ij} = \sum_{i=1}^{3}(\alpha\gamma)_{ik} = \sum_{j=1}^{3}(\alpha\beta)_{ij} = \sum_{k=1}^{3}(\alpha\gamma)_{ik} = \sum_{j=1}^{3}(\beta\gamma)_{jk} = \sum_{k=1}^{3}(\beta\gamma)_{jk} = 0$$

$$\sum_{i=1}^{3}(\alpha\beta\gamma)_{ijk} = \sum_{j=1}^{3}(\alpha\beta\gamma)_{ijk} = \sum_{k=1}^{3}(\alpha\beta\gamma)_{ijk} = 0 \tag{5.2}$$

Based on these zero-sum constraints, we can derive the following:

$$y_{ijkl} = \hat{\mu} + \hat{\alpha}_i + \hat{\beta}_j + \hat{\gamma}_k + \left(\widehat{\alpha\beta}\right)_{ij} + \left(\widehat{\alpha\gamma}\right)_{ik} + \left(\widehat{\beta\gamma}\right)_{jk} + \left(\widehat{\alpha\beta\gamma}\right)_{ijk} + r_{ijkl} \tag{5.3}$$

where

$$\hat{\mu} = \bar{y}_{....}$$

$$\hat{\alpha}_i = \bar{y}_{i...} - \bar{y}_{....}$$

$$\hat{\beta}_j = \bar{y}_{.j..} - \bar{y}_{....}$$

$$\hat{\gamma}_k = \bar{y}_{..k.} - \bar{y}_{....}$$

$$\left(\widehat{\alpha\beta}\right)_{ij} = \bar{y}_{ij..} - \bar{y}_{i...} - \bar{y}_{.j..} + \bar{y}_{....} \tag{5.4}$$

$$\left(\widehat{\alpha\gamma}\right)_{ik} = \bar{y}_{i.k.} - \bar{y}_{i...} - \bar{y}_{..k.} + \bar{y}_{....}$$

$$\left(\widehat{\beta\gamma}\right)_{jk} = \bar{y}_{.jk.} - \bar{y}_{.j..} - \bar{y}_{..k.} + \bar{y}_{....}$$

$$\left(\widehat{\alpha\beta\gamma}\right)_{ijk} = \bar{y}_{ijk.} - \bar{y}_{ij..} - \bar{y}_{i.k.} - \bar{y}_{.jk.} + \bar{y}_{i...} + \bar{y}_{.j..} + \bar{y}_{..k.} - \bar{y}_{....}$$

$$r_{ijkl} = y_{ijkl} - \bar{y}_{ijk.}$$

where the "." in the subscript means averaging over the levels of the corresponding factor or variable.

Now, the corrected total sum of squares and its components can be derived by subtracting $\bar{y}_{....}$ from both sides of equation (5.3), then squaring them, and calculating sums over i, j, k, l:

$$\sum_{i,j,k=1}^{3}\sum_{l=1}^{m}\left(y_{ijkl}-\bar{y}_{....}\right)^{2}=\sum_{i=1}^{3}3^{2}m\left(\hat{\alpha}_{i}\right)^{2}+\sum_{j=1}^{3}3^{2}m\left(\hat{\beta}_{j}\right)^{2}+\sum_{k=1}^{3}3^{2}m\left(\hat{\gamma}_{k}\right)^{2}$$

$$+\sum_{i,j=1}^{3}3m\left(\left(\widehat{\alpha\beta}\right)_{ij}\right)^{2}+\sum_{i,k=1}^{3}3m\left(\left(\widehat{\alpha\gamma}\right)_{ik}\right)^{2}+\sum_{j,k=1}^{3}3m\left(\left(\widehat{\beta\gamma}\right)_{jk}\right)^{2}$$

$$+\sum_{i,j,k=1}^{3}m\left(\left(\widehat{\alpha\beta\gamma}\right)_{ijk}\right)^{2}+\sum_{i,j,k=1}^{3}\sum_{l=1}^{m}\left(r_{ijkl}\right)^{2} \qquad (5.5)$$

The sum of squares of two-factor interactions can be divided into two components; for example, A × B can be split into AB and AB², each with two degrees of freedom.

To demonstrate this decomposition, we introduce two artificial factors, X and Z. Let the levels of A and B be denoted by i and j, respectively. Then, factors X and Z have factors s and t, respectively, determined as follows:

$$s = \text{mod}_3\left(i+j\right), \ t = \text{mod}_3\left(i+2j\right)$$

where mod₃ stands for "modulo 3"—in other words, a remainder of (.) after division by 3.

Table 5.2 shows how the levels s and t for the new effects X and Z vary throughout the experiment depending on the levels i and j of the two original factors A and B.

In the design shown, the four factors A, B, X, and Z are in balance in relation to each other and present a Graeco-Latin square (see Table 5.3).

TABLE 5.2

Levels s and t for Effects X and Z

i			j			$s = \text{mod}_3(i+j)$			$t = \text{mod}_3(i+2j)$		
0	1	2	0	0	0	0	1	2	0	1	2
0	1	2	1	1	1	1	2	0	2	0	1
0	1	2	2	2	2	2	0	1	1	2	0
	A			B			X			Z	

TABLE 5.3

Graeco-Latin Square Design

B＼A	0	1	2
0	$s=0, t=0$ (00)	$s=1, t=1$ (10)	$s=2, t=2$ (20)
1	$s=1, t=2$ (01)	$s=2, t=0$ (11)	$s=0, t=1$ (21)
2	$s=2, t=1$ (02)	$s=0, t=2$ (12)	$s=1, t=0$ (22)

This implies that the sum of squares for $s = 1,2,3$ and the sum of squares for $t = 1,2,3$ are orthogonal. Therefore, the AB interaction component represents the contrasts among $\bar{y}_{s=0}$, $\bar{y}_{s=1}$ and $\bar{y}_{s=2}$, and the AB^2 interaction component represents the contrasts among $\bar{y}_{t=0}$, $\bar{y}_{t=1}$, and $\bar{y}_{t=2}$. The sum of squares for these components can be calculated as follows:

$$SS_{AB} = 3m\left[\left(\bar{y}_{s=0} - \bar{y}_{.}\right)^2 + \left(\bar{y}_{s=1} - \bar{y}_{.}\right)^2 + \left(\bar{y}_{s=2} - \bar{y}_{.}\right)^2\right]; \bar{y}_{.} = \left(\bar{y}_{s=0} + \bar{y}_{s=1} + \bar{y}_{s=2}\right)/3$$

$$SS_{AB2} = 3m\left[\left(\bar{y}_{t=0} - \bar{y}_{.}\right)^2 + \left(\bar{y}_{t=1} - \bar{y}_{.}\right)^2 + \left(\bar{y}_{t=2} - \bar{y}_{.}\right)^2\right]; \bar{y}_{.} = \left(\bar{y}_{t=0} + \bar{y}_{t=1} + \bar{y}_{t=2}\right)/3$$

$$(5.6)$$

The A × B × C interaction has eight degrees of freedom and can be decomposed to ABC, ABC^2, AB^{2}C, and AB^{2}C^2. As before, let the levels of A, B, and C be denoted by i, j, and k, respectively. Thus, ABC, ABC^2, AB^{2}C, and AB^{2}C^2 represent the contrasts among the three groups of (i, j, k) satisfying each of the four systems of equations:

$$s = \mathrm{mod}_3\left(i+j+k\right),\, t = \mathrm{mod}_3\left(i+j+2k\right),\, u = \mathrm{mod}_3\left(i+2j+k\right),$$

$$v = \mathrm{mod}_3\left(i+2j+2k\right)$$

And the sums of squares of the components will be calculated as follows:

$$SS_{ABC} = 3m\left[\left(\bar{y}_{s=0} - \bar{y}_{.}\right)^2 + \left(\bar{y}_{s=1} - \bar{y}_{.}\right)^2 + \left(\bar{y}_{s=2} - \bar{y}_{.}\right)^2\right]; \bar{y}_{.} = \left(\bar{y}_{s=0} + \bar{y}_{s=1} + \bar{y}_{s=2}\right)/3$$

$$SS_{ABC2} = 3m\left[\left(\bar{y}_{t=0} - \bar{y}_{.}\right)^2 + \left(\bar{y}_{t=1} - \bar{y}_{.}\right)^2 + \left(\bar{y}_{t=2} - \bar{y}_{.}\right)^2\right]; \bar{y}_{.} = \left(\bar{y}_{t=0} + \bar{y}_{t=1} + \bar{y}_{t=2}\right)/3$$

$$SS_{AB2C} = 3m\left[\left(\bar{y}_{u=0} - \bar{y}_{.}\right)^2 + \left(\bar{y}_{u=1} - \bar{y}_{.}\right)^2 + \left(\bar{y}_{u=2} - \bar{y}_{.}\right)^2\right]; \bar{y}_{.} = \left(\bar{y}_{u=0} + \bar{y}_{u=1} + \bar{y}_{u=2}\right)/3$$

$$SS_{AB2C2} = 3m\left[\left(\bar{y}_{v=0} - \bar{y}_{.}\right)^2 + \left(\bar{y}_{v=1} - \bar{y}_{.}\right)^2 + \left(\bar{y}_{v=2} - \bar{y}_{.}\right)^2\right]; \bar{y}_{.} = \left(\bar{y}_{v=0} + \bar{y}_{v=1} + \bar{y}_{v=2}\right)/3$$

$$(5.7)$$

The ANOVA table with main effects, interactions, and their decompositions is presented in Table 5.4.

The mean squares are computed by dividing the sums of squares by their relating degrees of freedom. The residual mean square is an estimate of σ^2. The significance of a factor's main or interaction effect is tested using F statistics that are computed as dividing the corresponding mean squares by the residual mean square. The null hypothesis is that the particular effect equals zero. The corresponding F statistic has an F distribution where parameters are the degrees of freedom of the effect and of the residual mean square.

TABLE 5.4

ANOVA Table

Source of Variation	Degree of Freedom	Sum of Squares
A	2	$\displaystyle\sum_{i=1}^{3} 3^2 m\left(\hat{\alpha}_i\right)^2$
B	2	$\displaystyle\sum_{j=1}^{3} 3^2 m\left(\hat{\beta}_j\right)^2$
C	2	$\displaystyle\sum_{k=1}^{3} 3^2 m\left(\hat{\gamma}_k\right)^2$
A*B	4	$\displaystyle\sum_{i,\,j=1}^{3} 3m\left(\left(\widehat{\alpha\beta}\right)_{ij}\right)^2$
AB		$3m\left[\left(\bar{y}_{s=0}-\bar{y}_{.}\right)^2+\left(\bar{y}_{s=1}-\bar{y}_{.}\right)^2+\left(\bar{y}_{s=2}-\bar{y}_{.}\right)^2\right]$
AB²		$3m\left[\left(\bar{y}_{t=0}-\bar{y}_{.}\right)^2+\left(\bar{y}_{t=1}-\bar{y}_{.}\right)^2+\left(\bar{y}_{t=2}-\bar{y}_{.}\right)^2\right]$
A*C	4	$\displaystyle\sum_{i,\,k=1}^{3} 3m\left(\left(\widehat{\alpha\gamma}\right)_{ik}\right)^2$
B×C	4	$\displaystyle\sum_{j,\,k=1}^{3} 3m\left(\left(\widehat{\beta\gamma}\right)_{jk}\right)^2$
A×B×C	8	$\displaystyle\sum_{i,\,j,\,k=1}^{3} m\left(\left(\widehat{\alpha\beta\gamma}\right)_{ijk}\right)^2$
ABC	2	$3m\left[\left(\bar{y}_{s=0}-\bar{y}_{.}\right)^2+\left(\bar{y}_{s=1}-\bar{y}_{.}\right)^2+\left(\bar{y}_{s=2}-\bar{y}_{.}\right)^2\right]$
ABC²	2	$3m\left[\left(\bar{y}_{t=0}-\bar{y}_{.}\right)^2+\left(\bar{y}_{t=1}-\bar{y}_{.}\right)^2+\left(\bar{y}_{t=2}-\bar{y}_{.}\right)^2\right]$
AB²C	2	
AB²C²	2	
Residual	$27\,(m-1)$	$\displaystyle\sum_{i,\,j,\,k=1}^{3}\sum_{l=1}^{m}\left(y_{ijkl}-\bar{y}_{ijk.}\right)^2$
Total	$27m-1$	$\displaystyle\sum_{i,\,j,\,k=1}^{3}\sum_{l=1}^{m}\left(y_{ijkl}-\bar{y}_{...}\right)^2$

A full-factorial experiment allows estimating all main effects and all effects of the factor combinations. However, because the sample size grows exponentially with the number of factors, full-factorial designs are often too expensive to run. Also, since high-level interactions are often not active, it might be inefficient to execute such experiments. So, if the number of combinations in a full-factorial design is too high to be feasible, a fractional factorial design may be done, in which some of the combinations are omitted.

5.2.4 Fractional Factorial Experiment

Fractional factorial designs consist of a subset, or a fraction, of full-factorial design. As a consequence of using a fraction of the design, aliasing of factor effects and their combinations is created.

Let's consider a case when we have four factors A, B, C, and D, each on three levels. For the full-factorial experiment, a number of required runs will be $3^4 = 81$. Table 5.5 shows how the 80 degrees of freedom of this experiment would be allocated.

According to the effect hierarchy principle, one could expect that three-factor and four-factor interactions are not likely to be important. Table 5.5 shows that out of 80 degrees of freedom, 48 relate to such effects. It does not seem to be reasonable to run such an experiment.

The previously studied 3^3 full-factorial experiment employs a one-third fraction of the 3^4 design that is denoted as a 3^{4-1} design. This design is constructed by assigning the column for factor D to be equal to $\mod_3 (A + B + C)$. Such a relationship can be denoted as D = ABC.

As $i, j, k,$ and l represent the levels of factors A, B, C, and D, respectively, then

$$l = \mod_3 \left(i + j + k \right)$$

or equally

$$\mod_3 \left(i + j + k + 2l \right) = 0 \tag{5.8}$$

The last form of equation (5.8), in turn, means I = ABCD², which is called defining relation, and from which the aliasing patterns can be derived. For example, by adding $2i$ to both sides of equation (5.8), we have:

$$\mod_3 \left(3i + j + k + 2l \right) = 2i \rightarrow \mod_3 \left(j + k + 2l \right) = 2i$$

This means that A and BCD² are aliased. Following the same process, the following effects are aliased:

TABLE 5.5

Allocation of Degrees of Freedom for 3^4 Experiment

Main Effects	2-Factor Interactions	3-Factor Interactions	4-Factor Interactions
8	24	32	16

$$B = ACD^2 = AB^2CD^2, C = ABD^2 = ABC^2D^2, D = ABC = ABCD, AB = CD^2 = ABC^2D,$$

$$AB^2 = AC^2D = BC^2D, AC = BD^2 = AB^2CD, AC^2 = AB^2D = BC^2D^2, AD = AB^2C^2 = BCD,$$

$$AD^2 = BC = AB^2C^2D^2, BC^2 = AB^2D^2 = AC^2D^2, BD = AB^2C = ACD, CD = ABC^2 = ABD$$

$$(5.9)$$

A main effect or two-factor interaction is called *clear* if it is not aliased with any other main effect or two-factor interactions, and *strongly clear* if it is not aliased with any other main effect or two- or three-factor interaction. So, if three- and four-factor interactions are considered not important, then the aliasing relations (equation 5.9) demonstrate that A, B, C, D, AB^2, AC^2, AD, BC^2, BD, and CD can be estimated. These main effects or components of two-factor interactions are clear as they are not aliased with any other main effects or two-factor interaction components. While analyzing such an experiment, the ANOVA table will include only A, B, C, D, $AB = CD^2$, AB^2, $AC = BD^2$, AC^2, AD, $AD^2 = BC$, BC^2, BD, and CD effects.

For more details, we refer the readers to the vast existing literature on fractional design of experiments, for example, Montgomery (1991) or Wu and Hamada (2000, 2009).

5.2.5 Linear Mixed Models

A linear mixed model is a parametric model linear in the parameters which quantifies the relationships between a continuous dependent variable and independent variables that may involve a mix of fixed and random effects.

In general, fixed-effect parameters are associated with one or more continuous or categorical parameters and describe the relationships of the covariates to the dependent variable for an entire population. Fixed effects are unknown constant parameters, and estimation of these parameters is of fundamental interest, as they describe the relationships of the covariates with the continuous outcome variable. In experimental settings, for fixed-effect factors data are gathered from all the levels of the factor that are of interest.

Random effects are associated with one or more random factors and are specific to clusters or subjects within a population. When the levels of a factor can be thought of as having been sampled from a sample space, so that each particular level is not of specific interest, the effects associated with the levels of such factors can be modeled as random effects. In other words, random effects are directly used in modeling the random variation in the dependent variable at different levels of the data. In contrast to fixed effects, which are represented by constant parameters in a linear mixed model, random effects are represented by unobserved random variables.

The definition of the type of a factor and the selection of a model structure is critically important.

Suppose, as a result of our experiment, we gathered n observations of the response variable $y_1, \ldots, y_n$ and we want to estimate the impact of factors and their combinations on the response. The standard linear model can be formulated as follows:

$$y_i = \beta_0 + \beta_1 x_{i1} + \ldots + \beta_p x_{ip} + \varepsilon_i, \ i = 1, \ldots, n \tag{5.10}$$

where

y_i is the value of the response variable,

$x_{i1}, \ldots, x_{ip}$ are the fixed-effect regressors,

$\beta_0, \beta_1, \ldots, \beta_p$ are unknown fixed-effects parameters to be estimated,

ε_i are unknown independent and identically normally distributed random variables, $\varepsilon_i \sim N(0, \sigma^2)$.

The assumption that y_i are independently distributed is too restrictive. The mixed model extends the linear model (equation 5.10) by allowing responses to be correlated. Such an extension permits random effects and random coefficients in the analysis. The general linear mixed model is denoted as follows:

$$y_{ij} = \beta_0 + \beta_1 x_{ij1} + \ldots + \beta_p x_{ijp} + \gamma_{i1} z_{ij1} + \ldots + \gamma_{iq} z_{ijq} + \varepsilon_{ij}, \ i = 1, \ldots, n, \ j = 1, \ldots, m \tag{5.11}$$

where

y_{ij} is the value of the response variable for the observation i in group j,

$\beta_0, \beta_1, \ldots, \beta_p$ are unknown fixed-effects parameters,

$x_{ij1}, \ldots, x_{ijp}$ are the fixed-effect regressors for observation i in group j,

$\gamma_{ik}, k = 1, \ldots, m$ are unknown random variables, $\gamma_{ik} \sim N(0, \varphi_k^2)$, $\mathrm{Cov}(\varphi_k, \varphi_k') = \varphi_{kk}'$; φ_k^2 are the variances and φ_{kk}' the covariances among the random effects, assumed to be constant across groups,

$z_{ij1}, \ldots, z_{iq}$ are the random-effect regressors,

ε_{ij} are unknown random variables assumed to be multivariately normally distributed for group j.

In contrast to the standard linear model (equation 5.10), the residuals associated with observations in group j can be correlated.

The distinction between fixed and random factors, as well as their effects on a response variable, is critical in the context of linear mixed models. We refer the readers who are interested in studying the linear mixed models underlying theory to the extensive statistical literature on the topic, for example, Littell et al. (2007) and Verbeke and Molenberghs (1997).

5.2.6 Factors and Response Variables in the AI Framework

Let's consider the process of building a classifier using the design of experiment framework.

We may choose different approaches to create reasonable feature sets, but as a result, we end up with multiple possibilities. In the spirit of DoE, we call this a factor of feature sets. The number of feature set possibilities defines the number of levels of this factor. In the process of building the classifier, we need to make a decision which one of these feature sets, or, in other words, which level of the feature set factor, works best for our classification problem.

As the next step, we need to make up our mind about the machine learning method that we will use to build the classifier. Suppose we have reasons to choose a specific method. However, if this is not the case, then we have one more factor that can be named as a machine learning method. The number of levels of this factor is defined by the number of methods we want to try for building the classifier.

As discussed in Chapter 2, using machine learning methods requires assigning values to hyperparameters. For example, in the case of neural network, a decision should be made about a type of activation function, a number of hidden layers, neurons in each layer, etc. And in the case of support vector machine, we need to choose a kernel and its parameters, as well as the value of the cost parameter. Hyperparameters can be categorical (as an activation function or a type of kernel), integer (as a number of layers or degree of the polynomial kernel), or continuous (as a γ parameter of Gaussian kernel or a cost parameter). For categorical parameters, the number of levels is identified as a number of unique categorical values of the parameter. For integer parameters, the number of levels can be a number of unique integer values of the parameter, or, in case of many possible values, a number of intervals to which the values are grouped. In the case of continuous parameters, the values must be grouped into intervals, the number of which defines a number of levels.

An additional factor is a level of data contamination, where different ratios of data contamination define a number of this factor's levels.

5.2.7 Example

Let's say we would like to build a classifier for a dataset with 42 features, and suppose we decided to build a classifier using SVM with the sigmoid (hyperbolic tangent) kernel.

$$K(x,\ x') = \tanh\left(\alpha x^T x' + b\right) \tag{5.12}$$

where $\alpha > 0$ and $b < 0$.

The two hyperparameters of the sigmoid kernel are α and b. In some literature, it is suggested to use $\alpha = 1/N$, where N is the data dimension. In the case of 42 features, this value is 0.024, but we will use this value only as a starting point of our search of the values of the hyperparameters. As we do not have any reason to assume a linear relationship, we start with three levels for each factor. Let's set α on three levels defined as the following intervals: (0; 0.1], (0.1, 0.2], and (0.2, 0.3]; and the intercept b on three levels (−4; −3], (−3; −2], and (−2; −1].

Let's say we are looking for a classifier with the highest sensitivity (true positive rate). We need to find what combination of the kernel hyperparameters allows us to build a classifier that fits our criterion. In addition, we need to choose a value for cost variable C. As discussed in Chapter 2, for large values of cost value C, the SVM method will choose a smaller-margin hyperplane if that hyperplane does a better job of getting all the training points classified correctly. On the other hand, a very small value of C will cause the optimization part of SVM to look for a larger-margin separating hyperplane, even if that hyperplane misclassifies more points. Controlling for wider- or narrower-margin hyperplane allows addressing the problem of bias–variance trade-off. Let's assign three levels to the cost variable as (10; 100], (100; 300], and (300; 600].

The values of parameters in intervals associated with a specific level are given by random selection according to uniform distribution from the interval.

As discussed in Chapter 3, we create multiple training datasets to increase data variability, and for this example, let's use 1,000 training datasets.

So, our experiment has three factors: hyperparameters α and b and cost parameter C, all on three levels, 1,000 training datasets; and sensitivity is the response variable. Thus, the design presented in Table 5.1 works for this example, where factor A stands for the hyperparameter α, factor B stands for the hyperparameter b, and factor C stands for the cost parameter C. Each run of the experiment is randomly assigned to the training datasets, creating 37 replications per run. Execution of each run of this experiment includes the following steps:

1. Calculate the values of hyperparameters by a random selection of a value from an interval defined by the factor level. Let's say we execute run number 15 (see Table 5.1), where factor A is on level 1, factor B is on level 1, and factor C is on level 2. Then, the hyperparameter α is randomly selected from the interval (0.1, 0.2], the hyperparameter b is randomly selected from the interval (−3; −2], and the cost parameter C is randomly selected from the interval (300; 600].

2. Estimate the classifier model using SVM with the sigmoid (hyperbolic tangent) kernel and the set of hyperparameters defined in Step 1 for each replication in each run.

3. For each classifier, calculate sensitivity. In the experiment of this example, we will receive 37 values of sensitivity for each run.

As a next step, the results of this experiment need to be analyzed using ANOVA and linear mixed model approach.

5.2.8 Analysis of Linear Mixed Model Using SAS

Littell et al. (2007) is probably the best source describing SAS procedures for the estimation of mixed models. The MIXED procedure is the one that we use in the AI framework, as it fits a variety of mixed linear models and enables using these fitted models to make statistical inferences about the data. The CLASS statement of the MIXED procedure names the classification variables to be used in the model. The MODEL statement names a single dependent variable and the fixed effects, which determine the matrix of the mixed model. The RANDOM statement defines the random effects of the mixed model and can be used to specify variance component models as well as random coefficients. The random effects can be classification, and then defined in the CLASS statement, or continuous. The LSMEANS statement computes least squares means of fixed effects.

An example described in 5.2.7 can be analyzed using PROC MIXED.

```
proc mixed data=svm_experiment;
    class ai bi ci a(ref=first) b(ref=first) c(ref=first);
    model y_resp = a b c a*b a*c b*c / solution;
    random ai(a) bi(b) ci(c);
run;
```

This PROC MIXED estimates the model of three factors A, B, and C and their second-degree combinations. This model also takes into consideration that the actual values of hyperparameters are randomly selected from intervals associated with the factor levels. Thus, Ai, Bi, and Ci are random factors defining the actual values of each hyperparameter. The results of the analysis appear in Table 5.6, from which we can infer that second-degree combinations are not statistically significant in this experiment.

The code below estimates the model without second-degree interactions and creates svm_pred dataset with predicted response values.

```
proc mixed data= svm_experiment;
    class ai bi ci a(ref=first) b(ref=first) c(ref=first) ;
    model y_resp = a b c /  solution outp=svm_pred;
    random ai(a) bi(b) ci(c) ;
run;
```

The results of the fixed-effects estimation are presented in Table 5.7.

Using the predicted values presented in Table 5.8, we can reveal the sets of hyperparameters that lead to the best response values of sensitivity.

The top three conditions are as follows:

- A on level 2, B on level 1, and C on level 0 with an estimated value of sensitivity equal to 0.95

TABLE 5.6

Solution for Fixed Effects

Effect	A	B	C	Estimate	Std. Error	DF	t Value	Pr> |t|
Intercept				0.5130	0.006677	980	76.82	<.0001
A	1			0.09227	0.008390	980	11.00	<.0001
A	2			0.3411	0.008390	980	40.66	<.0001
A	0			0	.	.	.	.
B		1		0.08756	0.008390	980	10.44	<.0001
B		2		0.05462	0.008390	980	6.51	<.0001
B		0		0	.	.	.	.
C			1	−0.1781	0.008390	980	−21.23	<.0001
C			2	−0.4680	0.008390	980	−55.78	<.0001
C			0	0	.	.	.	.
A*B	1	1		−0.00522	0.009191	980	−0.57	0.5702
A*B	1	2		−0.00579	0.009191	980	−0.63	0.5286
A*B	1	0		0	.	.	.	.
A*B	2	1		0.01015	0.009191	980	1.10	0.2696
A*B	2	2		0.002005	0.009191	980	0.22	0.8274
A*B	2	0		0	.	.	.	.
A*B	0	1		0	.	.	.	.
A*B	0	2		0	.	.	.	.
A*B	0	0		0	.	.	.	.
A*C	1		1	0.003928	0.009191	980	0.43	0.6692
A*C	1		2	0.006448	0.009191	980	0.70	0.4831
A*C	1		0	0	.	.	.	.
A*C	2		1	−0.00891	0.009191	980	−0.97	0.3325
A*C	2		2	0.009728	0.009191	980	1.06	0.2901
A*C	2		0	0	.	.	.	.
A*C	0		1	0	.	.	.	.
A*C	0		2	0	.	.	.	.
A*C	0		0	0	.	.	.	.
B*C		1	1	0.01244	0.009191	980	1.35	0.1761
B*C		1	2	−0.00903	0.009191	980	−0.98	0.3258
B*C		1	0	0	.	.	.	.
B*C		2	1	0.003157	0.009191	980	0.34	0.7313
B*C		2	2	−0.00727	0.009191	980	−0.79	0.4289
B*C		2	0	0	.	.	.	.
B*C		0	1	0	.	.	.	.
B*C		0	2	0	.	.	.	.
B*C		0	0	0	.	.	.	.

TABLE 5.7

Solution for Fixed Effects, Step 2

Effect	A	B	C	Estimate	Std. Error	DF	t Value	Pr> |t|
Intercept				0.5117	0.004318	72	118.51	<.0001
A	1			0.09206	0.003791	108	24.29	<.0001
A	2			0.3454	0.003791	108	91.13	<.0001
A	0			0	.	.	.	.
B		1		0.09034	0.003823	72	23.63	<.0001
B		2		0.05198	0.003823	72	13.60	<.0001
B		0		0	.	.	.	.
C			1	−0.1746	0.004158	72	−41.99	<.0001
C			2	−0.4680	0.004158	72	−112.57	<.0001
C			0	0	.	.	.	.

TABLE 5.8

Predicted Values

A	B	C	N Obs.	Mean	Lower CI	Upper CI
0	0	0	37	0.5116676	0.508475	0.514861
0	0	1	37	0.3370877	0.333895	0.340281
0	0	2	37	0.0436354	0.040442	0.046828
0	1	0	37	0.6020080	0.598815	0.605201
0	1	1	37	0.4274282	0.424235	0.430621
0	1	2	37	0.1339759	0.130783	0.137169
0	2	0	37	0.5636498	0.560457	0.566843
0	2	1	37	0.3890699	0.385877	0.392263
0	2	2	37	0.0956176	0.092425	0.098811
1	0	0	37	0.6037231	0.60053	0.606916
1	0	1	37	0.4291432	0.42595	0.432336
1	0	2	37	0.1356909	0.132498	0.138884
1	1	0	37	0.6940635	0.690871	0.697256
1	1	1	37	0.5194837	0.516291	0.522677
1	1	2	37	0.2260314	0.222838	0.229224
1	2	0	37	0.6557053	0.652512	0.658898
1	2	1	37	0.4811254	0.477932	0.484318
1	2	2	37	0.1876731	0.18448	0.190866
2	0	0	37	0.8571066	0.853914	0.8603
2	0	1	37	0.6825267	0.679334	0.68572
2	0	2	37	0.3890745	0.385882	0.392267
2	1	0	37	0.9474471	0.944254	0.95064
2	1	1	37	0.7728672	0.769674	0.77606
2	1	2	37	0.4794149	0.476222	0.482608
2	2	0	37	0.9090888	0.905896	0.912282
2	2	1	37	0.7345089	0.731316	0.737702
2	2	2	37	0.4410566	0.437864	0.44425

- A on level 2, B on level 2, and C on level 0 with an estimated value of sensitivity equal to 0.91, and

- A on level 2, B on level 0, and C on level 0 with an estimated value of sensitivity equal to 0.86.

Thus, when building a classifier using SVM with the sigmoid kernel, the best sensitivity can be achieved with the α hyperparameter from the interval (0.2, 0.3], the b hyperparameter from the interval (−3; −2], and the cost variable from the interval (10; 100].

In this example of an experiment, we did not consider the values of specificity. So, by extending this experiment to measuring two response variables, sensitivity and specificity, we may identify different winners.

5.2.9 Analysis of Linear Mixed Model Using R

The "lme4" R package was developed by the team of Douglas Bates, Martin Maechler, Ben Bolker, and Steven Walker, and it provides functions to fit and analyze linear mixed models, generalized linear mixed models, and nonlinear mixed models. An alternative to "lme4" for mixed modeling in R is the "nlme" package authored by José Pinheiro and Douglas Bates. The main differences between the "nlme" and "lme4" packages are as follows:

- "lme4" uses modern, efficient linear algebra methods from the "eigen" package and uses reference classes, and thus, it is faster and more memory-efficient than "nlme" while using large objects.

- "lme4" includes generalized linear mixed model capabilities leveraging the "glmer" function.

- "lme4" does not currently implement features for modeling heteroscedasticity and correlation of residuals, which are implemented in "nlme."

- "lme4" offers built-in features for likelihood profiling and parametric bootstrapping.

- "lme4" is more modular than "nlme," and its components are easier to reuse for extensions of the basic mixed model framework.

In our framework, we use the "lme4" package that provides functions for fitting and analyzing mixed models, "lmer" for linear models, "glmer" for generalized linear models, and "nlmer" for nonlinear models. The parameters of these functions are described in Table 5.9.

The selection of what function to use depends on the structure of the model. For the example in 5.2.7, "lmer" can be successfully applied.

TABLE 5.9

Arguments of the "lmer," "glmer," and "nlmer" Functions

	Argument	Functions	Description
1	formula	lmer, glmer, nlmer	A two-sided linear formula object describing both the fixed-effects and random-effects parts of the model; for nlmer, it is a three-part "nonlinear mixed model" formula
2	data	lmer, glmer, nlmer	An optional data frame containing the variables named in the formula
3	family	glmer	A GLM family
4	REML	lmer	A logical parameter—to choose the REML criterion vs. log-likelihood
5	control	lmer, glmer, nlmer	A list containing control parameters, including the nonlinear optimizer to be used and parameters to be passed through to the nonlinear optimizer
6	start	lmer, glmer, nlmer	A named list of starting values for the parameters in the model, or a numeric vector
7	nAGQ	glmer, nlmer	The number of points per axis for evaluating the adaptive Gauss–Hermite approximation to the log-likelihood
8	subset	lmer, glmer, nlmer	An optional expression indicating the subset of the rows of data that should be used in the fit
9	weights	glmer, nlmer	An optional vector of "prior weights" to be used in the fitting process
10	offset	glmer, nlmer	An a priori known component to be included in the linear predictor during fitting
11	mustart	glmer	Optional starting values on the scale of the conditional mean

5.3 Summary

The selection of a set of hyperparameters is a process that is far from being intuitive or allowing some "rule-of-thumb" approach. At the same time, the impact of hyperparameter values on the quality of classifiers estimated with such hyperparameters is tremendous. There are not only an infinite number of values of each hyperparameter, but what is more important, there are also an infinite number of combinations of values of hyperparameters that impact the quality of classifiers. Without a systematic approach, to find sets of hyperparameters that lead to better classifiers is virtually impossible. In this chapter, we demonstrated how the design of experiment statistical methodology is used as a framework for optimizing values of hyperparameters, with the objective to find the best setup for a classifier. Analysis of the experimental results is performed by estimating a linear mixed model. We also reviewed the functions available in SAS and R to design an experiment and to analyze its results.

References

Box, G. E. P., Hunter, W. G., and Hunter, J. S. 1978. *Statistics for Experimenters: An Introduction to Design, Data Analysis, and Model Building.* New York: Wiley.

Cochran, W. G., and Cox, G. M. 1997. *Experimental Designs.* 2nd Ed. New York: John Wiley & Sons.

Cox, D. R. 1958. *Planning of Experiments.* New York: Wiley.

Dean, A. M., and Voss, D. 1999. *Design and Analysis of Experiments.* New York: Springer.

Dong, G., and Liu, H., ed. 2018. *Feature Engineering for Machine Learning and Data Analytics.* Boca Raton, Florida: CRC Press.

Hicks, C. R., and Turner, K. V. 1999. *Fundamental Concepts in the Design of Experiments.* 5th Ed. New York: Oxford University Press.

Hinkelmann, K., and Kempthorne, O. 1994. *Design and Analysis of Experiments.* New York: Wiley.

Kempthorne, O. 1952. *The Design and Analysis of Experiments.* New York: Wiley; London: Chapman & Hall.

Littell, R. C., Milliken, G. A., Stroup, W. W., Wolfinger, R. D., and Schabenberger, O. 2007. *SAS for Mixed Models.* North Carolina: SAS Institute.

Montgomery, D. C. 1991. *Design and Analysis of Experiments.* 3rd Ed. New York: Wiley.

Steinberg, D. M., and Hunter, W. G. 1984. Experimental design: Review and comment. *Technometrics*, 26 (2): 71–97.

Verbeke, G., and Molenberghs, G. 1997. *Linear Mixed Models in Practice. A SAS-Oriented Approach.* New York: Springer.

Weston, J., Mukherjee, S., Chapelle, O., Pontil, M., Poggio, T., and Vapnik, V. 2000. Feature selection for SVMs. *Advances in Neural Information Processing Systems*, 13. Neural Information Processing Systems Foundation, Inc.

Wu, C. F. J., and Hamada, M. 2000. *Experiments: Planning, Analysis, and Parameter Design Optimization.* New York: Wiley.

Wu, C. F. J., and Hamada, M. S. 2009. *Experiments: Planning, Analysis, and Optimization.* New York: John Wiley& Sons.

Part II

Part II

6

Introduction to the SAS- and R-Based Table-Driven Environment

6.1 Principles of Code-Free Design

The massive proliferation of the SAS and R has created new opportunities and new challenges for analytical applications designers and programmers. The user expectations for new analytical applications include short delivery lead times, ease of use, increased flexibility, and ability to scale. In response to these challenges, application designers and programmers are adopting rapid application development techniques. Code-free design is one of these techniques (Kolosova and Berestizhevsky, 1995). Code-free design lets the application designer describe applications visually, in terms of their functionality, and permits the development and maintenance of large-scale analytical applications without the writing of application computer code. The code is generated directly from definitions of application objects and actions to be applied to these objects. These definitions are created and stored in specially structured tables that comprise a table-driven environment, and the role of the application programmer is changed.

The main principles of code-free design are as follows:

- Application design is perceived as data and nothing but data. This means that the application design is defined in a set of specially structured tables and is stored, updated, and managed in the same way as ordinary data.

- Application activities are stated in terms of what must be done, but not how to do it. An application activity can be imagined as a definition of states, and messages allowing transition from one state to another. Of course, the states and messages are stored, updated, and managed as ordinary data. The application designer defines full application functionality by the use of its states and messages.

- The application is managed from a single control point.

To automate an application development process from beginning to end means that:

- requirements feed seamlessly into the design,
- design automatically writes code,
- code is deployed automatically into production, and
- maintenance changes automatically rewrite code.

Code-free design can realize this end-to-end automation through the table-driven environment. The table-driven environment is the most fundamental aspect of the code-free approach. The heart of this environment is the set of specially structured tables forming the data dictionary. The data dictionary contains a variety of information regarding application objects and operations, such as data structures and application activities. In other words, the data dictionary contains application design, and the data dictionary programs transform it into appropriate, functional working applications. The data dictionary also supports modifications of the application according to changes in application design. Different structures of data dictionaries can be defined depending on the application needs.

The table-driven environment consists of tables and computer programs that generate SAS or R code from the data stored in these tables. The table-driven environment provides a convenient and reliable framework for code-free application design.

When we look at an analytical application, we can see two important components: objects and operations. Each entity of the real world can be represented as an object. Consider the example of the application that supports marketing experiments. In this example, the objects are an audience, marketing channel, creative, and so on. In the real world, objects are continuously being transformed. These transformations are called operations. Only with a good understanding of these two components—objects and operations—can we build effective analytical applications.

Relational technology and the concept of code-free approach are two fundamental things that help to make generalizations about objects.

6.2 The Data Dictionary Components for the AI Framework

6.2.1 Relational Model

The relational data model consists of the following elements:

$$Model = \{S, C, O\}$$

where

- *S* is a set of rules that define the data structure: the objects, their properties, and the relationships between the objects.
- *C* is a set of rules that define the constraints that are applied to the defined data structure.
- *O* is a set of operations that are given to perform on the defined data structure.

Without a model, entity relationships are hidden in program code. The model brings these relationships to the surface. Also, the model helps ensure that programmed representations and interactions among the objects are valid. During application design, the relational data model allows the designer to represent the exact structure and properties of the data required for the application.

6.2.2 Table

A table is a fundamental entity in the relational model. A table can be represented as a two-dimensional object arranged in rows and columns. The structure of a table is very similar to that of an SAS dataset or R data frame.

6.2.3 Data Aspects

For a variety of technical and non-technical reasons, the relational model has proven to be very popular. Because the use of the relational model is so widespread and because it satisfies the functional requirements of many analytical applications, we have chosen to use a relational model in the design of the table-driven environment. The relational data model addresses three aspects of data:

- data structure
- data manipulation
- data integrity.

Data structure terms and their definitions are given below.

6.2.4 Relational Data Structure

The relational data model is responsible for the definition of the form in which data is represented (table structures) and for the definition of a set of operators that are enabled for these table structures. The table structures, together with the operators allowed on these tables, compose the relational data structure.

6.2.5 Domains

A domain is the set of all possible data values of a particular type for each column in a table. The domain of a column can also be defined in terms of another column; that is, the first column may contain only values from another column. The significance of domains is as follows:

- If two columns draw their values from the same domain, then comparisons involving these two columns make sense because they are comparing like values.
- If two columns draw their values from different domains, then comparisons involving these two columns do not make sense.

Domains may or may not be explicitly stored as actual sets of values. But they should be specified as part of the application definition, and then each column definition should include a reference to the corresponding domain. A given column can have the same name as the corresponding domain, or a different name.

6.2.6 Relations and Tables

The relation is an important term in the relational data model. Relations describe the connection between columns. The number of rows in a relation is called the cardinality of the relation. Relations can be conveniently represented in tabular form. A relational table is a data structure consisting of rows and columns, with each row being unique to that table. A table and a relation are not the same things, though for the purposes of our book we have assumed that they are. The relational data model gives symbolic names to data, thereby enabling to manipulate the data in the logical form. In the relational data model, each table has a unique name that identifies it. Each column in the table has a unique name (within this table) that identifies the column. The same column name can appear in two different tables. In this case, we can identify a column in one equivalent way by joining the table and column names. Throughout this book, we often refer to the column of the table as <Table name>.<COLUMN NAME>.

6.2.7 Functions

The function is the specific type of relationship between two relational tables. Suppose

- that each row of the first table relates exactly to each row of the second table, or
- that more than one row of the first table relates to one row of the second table, or
- that rows of the first table do not relate to rows of the second table.

Each of these relationships is called a function between the first table and the second table. These functions define the relationship or mapping between tables. Relationships between tables in the relational data model can be one-to-one (1:1) or one-to-many (1:M).

6.2.8 One-to-one Relationship

A one-to-one relationship (1:1) is the simple relationship that exists between tables in which each row of the first table corresponds to one row of the second table, and vice versa.

6.2.9 One-to-many Relationship

A one-to-many relationship (1:M) is the relationship that exists between tables in which each row of the first table corresponds to one row of the second table, but not vice versa.

6.2.10 Primary Key

According to the relational data model, every table must have a primary key. A primary key is a column, or set of columns, that makes each row of the table unique. Any column (or columns) can be used as the primary key as long as it defines a unique value for every row in the relational table.

6.2.11 Foreign Key

Another term that is used in the relational data model is a foreign key. Foreign keys are used to "link" related tables together. A foreign key is a column, or set of columns, that defines a data value that must exist in some other relational table as that other table's primary key. A relational data model maintains relational "links" by matching primary key values in one table with identical foreign key values in a related table. Note that a table can only have one primary key, but can have any number of foreign keys (since one table can be related to any number of other tables).

6.2.12 Missing Values

The concept of missing value is necessary because real-world information is frequently incomplete, and we need some way of handling such incompleteness. Missing values are useful when:

- a new row is to be created, but the user is currently unable to supply values for certain columns. An application can supply missing values for the new row in those columns.

- a new column is to be added to an existing table. The value of the new column can be defined as the missing value in all existing rows.
- an aggregate function, such as a summary, is to be applied to a column of a table. In computing the function, it is desirable for an application to recognize and ignore rows for which missing values have been supplied. (It is also desirable for users to understand what is going on in such a situation; for example, the average is not equal to the sum divided by the count.)

To represent a missing value of a column in the table, it is necessary to find a symbol that is different from all legal symbols in this column.

6.2.13 Data Dictionary

Everything about an application is defined in specially structured tables that comprise a table-driven environment. These tables are sophisticated and allow a great deal of complexity in the final application. The special programs transform data from these tables (metadata) into application code. Such tables and programs form a data dictionary.

6.3 Properties of the Data Dictionary

The data dictionary is the repository of a variety of information regarding objects related to an application, such as application tables and their columns, users, access privileges, and integrity constraints. In order to define an application, we need to build the set of data dictionary tables that will contain data (metadata) about the application's objects.

The possible set of data dictionary tables includes six tables: Library, Object, Location, Message, Property, and Link. This set is called the Kernel set and may be extended by additional tables at any time. The data dictionary tables themselves can be defined in the Kernel set and handled by the data dictionary like any other application.

Below we describe the tables of the Kernel set in terms of the relational data model.

6.3.1 The Library Table

The Library table lists SAS libraries specifications about data locations. These specifications of data locations can be used by R programs as well. The columns of the Library table are defined as shown in Table 6.1.

The LIBRARY column is the primary key of the Library table.

TABLE 6.1

Columns of the Library Table

Column Name	Type	Length	Description
LIBRARY	Character	8	SAS library name
LOCATION	Character	80	Operating system-specific file name for SAS library

6.3.2 The Object Table

The next table is the Object table, which lists application tables and the names of corresponding SAS datasets or R datasets where their data is stored. The Object table also defines application table for SAS datasets. The columns of the Object table are defined as shown in Table 6.2.

The TABLE column is the primary key of the Object table.

6.3.3 The Location Table

The Location table lists application tables and SAS libraries, where SAS datasets corresponding to the tables are kept. The columns of the Location table are defined in Table 6.3.

The TABLE and LIBRARY columns together form the primary key of the Location table.

The Location and Object tables together define physical patterns (datasets) for each application table.

6.3.4 The Message Table

The next is the Message table that lists application messages. The columns of the Message table are defined as shown in Table 6.4.

The MESSAGE column is the primary key of the Message table.

TABLE 6.2

Columns of the Object Table

Column Name	Type	Length	Description
TABLE	Character	8	Table name
TITLE	Character	80	Table title
DATASET	Character	8	SAS or R dataset name where data of the table is stored

TABLE 6.3

Columns of the Location Table

Column Name	Type	Length	Description
TABLE	Character	8	Table name
LIBRARY	Character	8	SAS library name

TABLE 6.4

Columns of the Message Table

Column Name	Type	Length	Description
MESSAGE	Numeric	8	Message identification number
TEXT	Character	80	Message text
MESTYPE	Character	1	Message type: E—error message, W—warning, I—informative message
OUTPUT	Character	2	Output message destination: S—screen and/or D—dataset

6.3.5 The Property Table

The Property table specifies the properties of application tables. The Property table lists the columns of each application table. For each column, the following information is defined:

- name of the table of which that column is a part
- column name
- column data type
- column length
- column domain
- column location on the screen form, etc.

The columns of the Property table are defined in Table 6.5.

TABLE 6.5

Columns of the Property Table

Column Name	Type	Length	Description
TABLE	Character	8	Table name
COLUMN	Character	8	Column name
TITLE	Character	80	Column title
TYPE	Character	1	Column data type: C—character or N—numeric
LENGTH	Numeric	8	Column length
ATTRIBUT	Character	2	Column property: P—primary key
DOMTAB	Character	8	Domain table name
DOMCOL	Character	8	Domain column name
MEANTAB	Character	8	Meaning table name
MEANCOL	Character	8	Meaning column name
INITVAL	Character	80	Initial column's value
FORMULA	Character	80	The formula for the computed column
MISSING	Character	1	Code of missing value
MESSAGE	Numeric	8	Error message identification number

The TABLE and COLUMN columns together form the primary key of the Property table.

Table 6.6 shows the example of the properties of the Message table defined in the Property table.

6.3.6 Meaning

A "meaning" is a simple way to replace an uninformative value with meaningful text. The definition of the meaning for a column is merely a reference to the application table and its column that contains a meaningful description. The correct definition of the meaning is based on the relationships between application tables. The column for which we define meaning must be the foreign key of the table that contains meaningful description.

6.3.7 The Link Table

The Link table specifies pairs of tables linked by foreign keys. The columns of the Link table are defined as shown in Table 6.7.

The TABLE, COLUMN, RELTABLE, and RELCOL columns together form the primary key of the Link table.

6.3.8 Process of Application Data Model Definition

It is important to mention that application data model definition can be performed with iterations, and it is not necessary to complete the application design process before using the application, nor is it necessary to get

TABLE 6.6

Property Table (Selected Columns)

TABLE	COLUMN	TYPE	LENGTH	ATTRIBUT	INITVAL
Message	MESSAGE	N	8	P	
Message	TEXT	C	80		
Message	MESTYPE	C	1		W
Message	OUTPUT	C	2		S
...	...	...	...	...	...

TABLE 6.7

Columns of the Link Table

Column Name	Type	Length	Description
TABLE	Character	8	Table name
RELTABLE	Character	8	Name of the related table
COLUMN	Character	8	Column name
RELCOL	Character	8	Name of the column from the related table

everything right the first time. This does not mean that application design is unnecessary in a table-driven development environment; application design is still needed. However,

- it does not all have to be done at once;
- it does not have to be perfect the first time; and
- if requirements change, then the design can be changed too, in a relatively painless manner.

The most ambitious goal of data dictionary-based development is to maximize the quality of application components while minimizing their development and maintenance costs.

6.3.9 Features of the Data Dictionary

There are features of the data dictionary that constitute a checklist of functionality:

Relational: The data dictionary should be based on relational technology. This is the current standard in the software industry.

Extendable: The data dictionary should be extensible. The designer should have the ability to extend the data dictionary as required, to add new files, databases, and other related information sources. In other words, the data dictionary's metamodel, or data about data, should be flexible enough to be extended as required.

Meta-meta model: The data dictionary should be based on an object-oriented meta-meta model. The meta-meta model can be thought of as a data dictionary for the data dictionary. A data dictionary constructed in this manner is self-defining and self-maintaining. It is self-defining in that just as a designer uses the relational model and data dictionary to define metadata, the data dictionary is defined using the same entity–relationship model (meta-metadata). Development of this meta-meta model in an object-oriented, extensible manner ensures that future growth of the application is consistent, logical, and responsive to desires.

Preexisting data definitions: The data dictionary should provide the ability to populate the data dictionary with preexisting data definitions automatically. Since most corporate data preexists, the data dictionary must be able to generate data dictionary definitions from database descriptions automatically.

Database definitions: The data dictionary should provide the ability to generate database definitions automatically. Since it is nonproductive and potentially error-prone to permit programmers to self-code data descriptions, the data dictionary should provide this capability.

Common user access: The data dictionary should provide an easy-to-use, standard common user access interface. The data dictionary should provide a dialog that is object–action-oriented. Designers are presented with appropriate objects and all actions that can be invoked for the objects.

Report facility: The data dictionary should provide an integrated reporting facility.

Security: The data dictionary should have multiple levels of security that can be controlled by the data dictionary, data, or designer.

Life cycle support: The data dictionary should provide adequate facilities to track modifications of a data element's description as it progresses through the life cycle of planning, design, development, testing, and production.

6.3.10 The Components of the Optimization Framework and Their Definitions in the Data Dictionary

The possible set of data dictionary tables for the machine learning optimization framework includes five sets (components) of tables: input data, bootstrap, machine learning, experimental design, and training data contamination. We call these sets (components) the Framework set, and it can be extended by additional tables or set of tables at any time. Each of the components of the Framework set will be described in the next chapters.

6.4 Deployment of Code-Free Design with SAS and R

While the designer works with the contents of the data dictionary tables, the programmer works with the structures of those tables (table names, table columns, column data types, etc.). While the contents of the data dictionary tables can be continuously changed (in accordance with the changes in application requirements), their structures remain unchangeable. The data dictionary tables constitute the table-driven environment, and once developed, programs processing these tables generate a wide variety of applications.

The process of generating a table-driven environment is iterative by nature. The programmer must generate the tables which are the core of the data dictionary and then generate other data dictionary tables and application objects from the metadata supplied in these tables.

Below we demonstrate an SAS program that generates application objects from the metadata stored in a table-driven environment.

6.4.1 How to Generate Application Objects

The following %DATASET macro demonstrates how to generate SAS datasets from the metadata of the Object, Location, and Property tables.

```
/*
PROGRAM DATASET.SAS
DESCRIPTION Creates SAS data set for the specified table.
USAGE %dataset(libref=, table=, mis=) ;
PARAMETERS
libref - is the name of the library storing the data
dictionary data sets.
table - is the name of the table for which is required to
create the SAS data set.
mis - is the code identifying the missing value.
REQUIRES The object, property, location, library data sets
must exist.
AUTHORS T.Kolosova and S.Berestizhevsky
*/
%macro dataset (libref=, table=, mis=) ;
/*
The following DATA step creates dataset macro variable
containing the name of data set corresponding to the specified
table.
*/
    data _null_ ;
        set &libref..Object ;
        where upcase(left(table)) = upcase("&table") ;
        call symput("dataset", trim(dataset)) ;
    run ;
    %let count = 0 ;
    %let _icount = 0 ;
/*
The following DATA step creates macro variables and fills them
with data from the property data set:
count - contains number of columns of the specified table.
_v0 - is a series of macro variables containing names of the
columns of the specified table with their types and length.
_v2 - is a series of macro variables containing names of the
columns of the specified table for which indexes are defined.
_v3 - is a series of macro variables containing names of the
columns of the specified table with their titles.
*/
    data _null_ ;
        retain _count 1 _icount 0 ;
        call symput("_v0" || left(_count), "&mis") ;
        set &libref..Property ;
        where upcase(left(table)) = upcase("&table") ;
        call symput("_v3" || left(_count), trim(left(column))
|| ' =
            %nrbquote(' || trim(left(title)) || ')') ;
        if index(upcase(attribut), "I") > 0 then
        do ;
            _icount + 1 ;
            call symput("_v2" || left(_icount), left(column)) ;
```

```
                    call symput("_icount", left(_icount)) ;
            end ;
            if upcase(type) = "C" then
                call symput("_v0" || left(_count),
                trim(left(column)) ||
                    " $" || left(length)) ;
            if upcase(type) = "N" then
                call symput("_v0" || left(_count),
                trim(left(column)) ||
                    " " || left(length)) ;
            call symput("count", _count) ;
            _count + 1 ;
    run ;
/*
The following DATA step generates SAS data set for the
specified table according to the values of the macro variables
created in the previous DATA step.
*/
    data &dataset ;
        length
        %do _j = 1 %to &count ;
            && _v0&_j
        %end ;
        ;
        label
        %do _j = 1 %to &count ;
            && _v3&_j
        %end ;
        ;
        stop ;
    run ;
/*
The following PROC SQL creates indexes, if any, for the SAS
data set generated in the previous DATA step according to the
values of the macro variables stored in the _v2 macro variable.
*/
    %if &_icount > 0 %then
    %do ;
        proc sql ;
        %if &_icount = 1 %then
        %do ;
            create index && v2&_icount on
            &dataset (&& v2&_icount) ;
        %end ;
        %else
        %do ;
            create index __index on &dataset
            ( && v21
            %do i = 2 %to &_icount ;
            , && v2&_icount
```

```
        %end ;
          ) ;
      %end ;
    %end ;
%mend dataset ;
```

For example, to generate the SAS dataset for the Demo table according to the metadata of the Object, Location, and Property tables, we have to invoke the %DATASET macro as follows:

```
%dataset(libref=kernel, table=Demo, mis=.);
```

6.4.2 Generating R Datasets from the Data Dictionary Metadata

The following DATASET R program demonstrates how to generate R datasets from the metadata of the Library, Object, and Property tables. These tables are stored in comma-separated values files.

```
#
# PROGRAM DATASET.R
# DESCRIPTION Creates R data set for the specified table.
# USAGE dataset(table, tmp_src, init_file)
# PARAMETERS table - is the name of the table for which is
# required to create the R data set.
# tmp_src - is the temporary name of file to store generated R
# code.
# init_file - is the name of the comma-separated value file
# that contains path to data dictionary files. This file is
# part of data dictionary and it is called library.csv
# object_file - is the name of comma-separated value file that
# contains meta-data about  application tables and data
# dictionary tables as well
# property_file - is the name of comma-separated value file
# that contains meta-data about  application tables structure
# and data dictionary tables structure as well
# REQUIRES The object.csv, property.csv files of data
# dictionary must exist.
# AUTHORS T.Kolosova and S.Berestizhevsky
#
dataset <- function (table, tmp_src, init_file, object_file,
property_file)
{
    lib<-function(start)
    {
        ds1<-read.csv(file = start, header=TRUE, sep=",")
        innerf <- function(text) assign("library", text, envir
        = .GlobalEnv)
        text <- paste(cat(sprintf('%s', ds1$location)))
```

```
        innerf(ds1$location)
    }
    loc<-function(object_file, property_file)
    {
        text<-paste0(as.character(library),"\\",
        as.character(object_file))
        assign("object", paste0(text), envir = .GlobalEnv)
        text<-paste0(as.character(library),"\\",
        as.character(property_file))
        assign("property", paste0(text), envir = .GlobalEnv)
    }
    create_ds<-function(table, tmp_src)
    {
        fn <- tmp_src
        if (file.exists(fn))
            file.remove(fn)
        columns <- vector()
        types <- vector()

        type <- "integer()"
        ds1<-read.csv(file = property, header=TRUE, sep=",")
        for (i in 1:nrow(ds1))
        {
            if (all(trimws(table) == ds1$table[i]) )
            {
                if (all("C" == ds1$type[i])) type
                <- "character()"
                columns <- c(columns, paste0(ds1$column[i],
                sep = " ", collapse = NULL))
                types <- c(types, paste0(type, sep = " ",
                collapse = NULL))
                type <- "integer()"
            }
        }
        sink(fn)
        text<- paste0(paste0(columns), '=', paste0(types),
        ',')
        text <-paste0(text, collapse=" ")
        text <- substr(text,1,nchar(text)-1)

        cat(sprintf('%s <- data.frame( %s )\n',  table,
        paste0(text) ))
        sink()
        source(fn)
    }

    lib(init_file)
    loc(object_file, property_file)
    create_ds(table, tmp_src)
}
```

```
dataset("Location", "tmp_source.txt", "c:\\datadictionary\\
library.csv", "\\object.csv", "\\property.csv")
```

6.4.3 SAS and R Interoperability

We describe one method of passing SAS data from SAS to R, using R to analyze data and produce results, and then passing the results back to SAS. The R package provides a wide range of machine learning functionality, some of which are unavailable, or time-consuming to achieve, in SAS. Using the method described below, this functionality can be made available to SAS applications. The basic principle of the method outlined below could also be adapted to create different data analysis processes using R.

Support vector machine (SVM) is a classification method that is not available in SAS/STAT. The package "e1071" in R provides the state-of-the-art implementation for SVM classification. We created a small SAS macro that implements interoperability between SAS and R. This SAS macro enables native R language to be embedded in and executed along with an SAS program in the SAS Base environment. This macro executes a user-defined R code in batch mode. Also, this macro automatically converts data between SAS datasets and R data frames such that the data and results from one statistical computing environment can be utilized by the other environment. The objective of this macro is to leverage the strength of the R within the SAS environment in a systematic manner.

```
/*
PROGRAM SVM.SAS
DESCRIPTION Support Vector Machine for Classification In SAS
By R
USAGE %%svm (train=, validate=, result=, targetvar=, tmppath=,
rpath=)
PARAMETERS train     = dataset for training
validate  = dataset for validation mis - is the code
identifying the missing value
result    = dataset created after prediction
targetvar = target variable
tmppath   = temporary path to store the exchange files between
SAS and R
rpath     = path for R executable
REQUIRES e1071 and foreign R packages should be installed
AUTHORS T.Kolosova and S.Berestizhevsky
*/

%macro svm (train = , validate = , result = , targetvar = ,
tmppath = , rpath = );
    proc export data = &train outfile = "&tmppath\sas2r_train.
    csv" replace;
    run ;
```

```
proc export data = &validate
    outfile = "&tmppath\sas2r_validate.csv" replace;
run;
proc sql;
    create table _tmp0 (string char(200));
    insert into _tmp0
    set string = 'train=read.csv("sas_path/sas2r_train.
    csv",header=T)'
    set string = 'validate=read.csv("sas_path/sas2r_
    validate.csv",header=T)'
    set string = 'require(e1071,quietly=T)'
    set string = 'model=svm(sas_targetvar ~ .
    ,data=train)'
    set string = 'predicted=predict(model,newdata=validate
    ,type="class")'
    set string = 'result=as.data.frame(predicted)'
    set string = 'require(foreign, quietly=T)'
    set string = 'write.foreign(result,"sas_path/r2sas_
    tmp.dat","sas_path/r2sas_tmp.sas",package="SAS")' ;
quit;

data _tmp1;
    set _tmp0;
    string = tranwrd(string, "sas_targetvar",
    propcase("&targetvar"));
    string = tranwrd(string, "sas_path",
    translate("&tmppath", "/", "\"));
run;

data _null_;
    set _tmp1;
    file "&tmppath\sas_r.r";
    put string;
run;

options xsync xwait;
x "R.exe CMD BATCH --vanilla --slave &tmppath\sas_r.r";

%include "&tmppath\r2sas_tmp.sas";

data &result;
    set &validate;
    set rdata;
run;

proc datasets nolist;
    delete _: rdata;
quit;
%mend;
```

```
%svm(train = iris, validate = iris, result = iris_result,
targetvar = species,  tmppath = c:\temp, rpath = C:\Progra~1\
R\R-3.5.3\bin\x64\R.exe);
```

6.5 Summary

This chapter introduced the table-driven environment that is used for the development of the AI framework. This environment supports the code-free design and facilitates fast and robust implementation. This chapter described tables, as well as SAS and R programs comprising the table-driven environment. In the next chapters, we demonstrate how to extend the table-driven environment to address such tasks as data input, bootstrap, design of experiments, and support vector machine classification.

Reference

Kolosova, T., and Berestizhevsky, S. 1995. *Table-Driven Strategies for Rapid SAS Applications Development.* Cary, NC: SAS Institute, Inc.

7

Input Data Component

7.1 Overview of Data Management

A data dictionary for the input data component is created to understand the meaning of input data (e.g., training and testing datasets) and define the processes of data cleansing and generation of multiple training and/or testing subsamples. The data dictionary for the input data component is always used during data analysis so that the meaning of variables and their values will never be a question.

Because input datasets are usually flawed, data cleansing is a very important step in the machine learning process. Data cleansing is the process used to identify inaccurate, incomplete, or improbable data and then to correct it when possible. To put it in a nutshell, data cleansing is a set of procedures that helps make your dataset more suitable for machine learning. To put it simply, the quality of training data determines the performance of the machine learning algorithms. Even if you think you have good-quality data, you can still run into problems with biases hidden within your training datasets. Thus, input data component is used to define data cleansing processes that include detection and treatment of outliers and bias corrections. Usually, data cleansing comprises of two stages: detection and correction, and we define these stages as metadata in the data dictionary for input data component.

7.1.1 Data Dictionary

7.1.1.1 The Input Data Dictionary

This data dictionary serves as a repository for metadata that defines input and output datasets and data cleansing and bootstrapping processes. This metadata then used by SAS macros or R-written programs to generate (execute) defined in the data dictionary processes such as the creation of output multiple datasets (subsamples of training and testing datasets) as well as perform cleansing and bootstrapping processes. Changes in the data

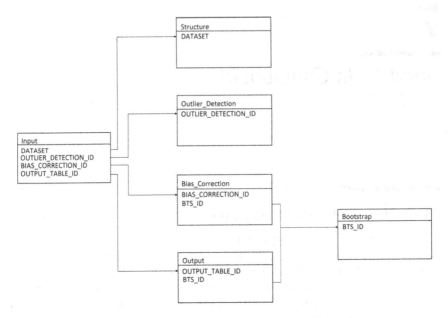

FIGURE 7.1
Data dictionary for input data component.

dictionary metadata will essentially change the needed processes to clean data and/or create training/testing datasets. The definition of processes as data (metadata) allows the execution of such processes without changing the source code of SAS macros and/or R-written programs.

A data dictionary for input data component includes, at a minimum, source dataset name, variable names, variable types, variable descriptions or labels, and output datasets (training and testing). It also includes definitions of data cleansing processes that in turn contain references to processes of outliers detection and treatment. There are six tables that comprise input data component data dictionary. They are Input, Structure, Outlier_Detection, Bias_ Correction, Bootstrap, and Output tables. Relations among the tables of the data dictionary for input data component are presented in Figure 7.1.

Below we provide a description of the structure of input data component tables along with examples of possible content.

7.1.1.2 Input and Structure Tables

The columns of the Input table are defined as shown in Table 7.1.

The INPUT_ID column comprises the primary key of the Input table. The Input table contains references to data cleansing processes to be applied to data defined in the table. These references are defined in the OUTLIER_ DETECTION_ID and BIAS_CORRECTION_ID columns, and these columns

TABLE 7.1

Columns of the Input Table

Column Name	Type	Length	Description
INPUT_ID	Character	8	Input ID
TITLE	Character	80	Input ID title
DATASET	Character	8	Dataset (file) name
OUTLIER_DETECTION_ID	Character	8	Reference to the primary key of the Outlier_Detection table
BIAS_CORRECTION_ID	Character	8	Reference to the primary key of the Bias_Correction table
OUTPUT_TABLE_ID	Character	2	Reference to the primary key of the Output table

comprise the primary keys of the Outlier_Detection and Bias_Correction tables, respectively. Those tables will be defined below.

The Structure table specifies the properties of datasets. The Structure table lists the columns of each dataset. For each column, the following information is defined: name of the dataset of which that column is a part, column name, column data type, column length, and column role—label or feature. Column role defines which column or columns are used to classify data using labels and which represent data features. The columns of the Structure table are defined as shown in Table 7.2.

The DATASET and VARIABLE_NAME columns comprise the primary key of the Structure table.

7.1.1.3 Outlier_Detection and Bias_Correction Tables

The Outlier_Detection and Bias_Correction tables define cleansing methods and their parameters. The columns of the Outlier_Detection table are defined as shown in Table 7.3.

TABLE 7.2

Columns of the Structure Table

Column Name	Type	Length	Description
DATASET	Character	8	Dataset name
VARIABLE_NAME	Character	8	Variable name
VARIABLE_TITLE	Character	80	Variable title
VARIABLE_TYPE	Character	1	Variable type: C—character or N—numeric
VARIABLE_LENGTH	Numeric	8	Variable length
VARIABLE_ROLE	Character	1	Variable role: F(eature) or L(abel)
FORMULA	Character	80	Formula for computed variable
MISSING	Character	1	Code of missing value

TABLE 7.3

Columns of the Outlier_Detection Table

Column Name	Type	Length	Description
OUTLIER_ DETECTION_ID	Character	8	Procedure ID
METHOD	Character	8	Method name (e.g., MCD—minimal covariance determinant method and columns H, CUTOFFALHA, etc., are intended to define parameters for MCD)
H	Numeric	8	Default value $(3*n+p+1)/4$, where n is the number of observations and p is the number of variables
CUTOFFALPHA	Numeric	8	Default value is 0.025
MCDALPHA	Numeric	8	Default value is 0.025

The OUTLIER_DETECTION_ID column comprises the primary key of the Outlier_Detection table.

The columns of the Bias_Correction table are defined as shown in Table 7.4.

The BIAS_CORRECTION_ID and SID columns comprise the primary key of the Bias_Correction table. The SID column lists sequential numbers to uniquely identify what bias correction procedure, defined in the Bootstrap table below and referenced via BTS_ID column, corresponds to which dataset variable that is defined in VAR_NAME column.

Values in BIAS_CORRECTION_ID and OUTLIER_DETECTION_ID columns listed in the Input table (see Table 7.1) associate the input dataset with bias correction and/or outliers detection processes defined in the Outlier_Detection and Bias_Correction tables. The latest contains a reference to the Bootstrap table.

7.1.1.4 Bootstrap Table

The columns of the Bootstrap table are defined as shown in Table 7.5.

TABLE 7.4

Columns of the Bias_Correction Table

Column Name	Type	Length	Description
BIAS_ CORRECTION_ID	Character	8	Bias correction ID
BIAS_TITLE	Character	8	Bias correction title
SID	Numeric	8	Sequential number to identify the corresponding BIAS_CORRECTION_ID with VAR_NAME, METRIC, and BTS_ID
VAR_NAME	Character	80	Name of the dataset variable
METRIC	Character	8	Metric: M(ean), M(e)D(ian)
BTS_ID	Character	8	Reference to the primary key of the Bootstrap table

TABLE 7.5

Columns of the Bootstrap Table

Column Name	Type	Length	Description
BTS_ID	Character	8	Bootstrap ID
BTS_TITLE	Character	8	Bootstrap title
RESAMPLE_SIZE_PCT	Numeric	8	For example, 100 means n-out-of-n; 70 or 50 means m-out-of-n
REPLICATIONS	Numeric		Number of subsamples to be created
DISTRIBUTION	Character	1	Name of the distribution for random sample selection: U(niform), where each observation has a uniform probability of $1/N$ of being selected at each step of the resampling process. Within the union of the B bootstrap samples, each observation has an *expected value* of appearing B times; B(alanced) where each observation appears *exactly* B times in the union of the B bootstrap samples of size N
STRATA_VAR	Character	32	Name of a variable that defines strata

The BTS_ID column comprises the primary key of the Bootstrap table. The data defined in the Bootstrap table identifies bootstrap processes that can be applied for bias correction as well as for the generation of multiple (this value is defined in the REPLICATIONS column) training or testing datasets. The STRATA_VAR defines a variable from the input dataset, the value of which should be used to separate the values of other variables in the dataset. The most applicable use of STRATA_VAR is to define a time series in the input dataset.

7.1.1.5 Output Table

The columns of the Output table are defined as shown in Table 7.6.

OUTPUT_TABLE_ID and OUTPUT_ROLE comprise the primary key of the Output table. The Output table contains a reference to the Bootstrap table.

TABLE 7.6

Columns of the Output Table

Column Name	Type	Length	Description
OUTPUT_TABLE_ID	Character	8	Output ID
OUTPUT_ROLE	Character	80	Output data role
TABLE_PREFIX	Character	8	
WITH_RETURN	Character	1	Select data from input dataset with or without return. Values are Y or N. Default is N
DATA_PCT	Numeric		Percent of input data used to create OUT_ID table
BTS_ID	Character	8	

To get a better understanding of what all the above-described metadata definitions mean, we created an example of input data component table pre-filled with metadata.

Tables 7.7–7.12 show the example of how the content of the data dictionary tables for input data component may look like.

TABLE 7.7

Content of the Input Table

INPUT_ID	TITLE	DATASET	OUTLIER_ DETECTION_ ID	BIAS_ CORRECTION_ ID	OUTPUT_ TABLE_ID
1	Title A	DataOne	1	2	1
2	Title B	DataTwo	2	1	2

TABLE 7.8

Content of the Structure Table

DATASET	VARIABLE_ NAME	VARIABLE_ TITLE	VARIABLE_ TYPE	VARIABLE_ LENGTH	VARIABLE_ ROLE	FORMULA	MISSING
DataOne	VarOne	Title One	Character	20	L	.	.
DataOne	VarTwo	Title Two	Numeric	8	F	.	.
DataOne	VarThree	Title Three	Numeric	8	F	.	.
DataOne	VarFour	Title Four	Numeric	8	F	VarTwo* VarThree/50	.
DataOne	VarFive	Title Five	Numeric	8	F	VarFour ** 3	.
DataTwo	VarOne	Title One	Character	20	L	.	.
DataTwo	VarTwo	Title Two	Numeric	8	F	.	.
DataTwo	VarSix	Title Six	Numeric	8	F	.	.

TABLE 7.9

Content of the Outlier_Detection Table

OUTLIER_ DETECTION_ID	METHOD	H	CUTOFFALPHA	MCDALPHA
1	MDC	.	0.025	0.05
2	MDC	100	0.01	0.05

TABLE 7.10

Content of the Bias_Correction Table

BIAS_ CORRECTION_ID	BIAS_TITLE	SID	VAR_NAME	METRIC	BTS_ID
1	Location	1	VarThree	Median	1
1	Location	2	VarFour	Mean	1

TABLE 7.11

Content of the Bootstrap Table

BTS_ID	BTS_TITLE	RESAMPLE_SIZE_PCT	REPLICATIONS	DISTRIBITION	STRATA_VAR
1	Title One	100	5,000	Uniform	VarOne
2	Title Two	70	10,000	Balanced	VarOne

TABLE 7.12

Content of the Output Table

OUTPUT_TABLE_ID	OUTPUT_ROLE	TABLE_PREFIX	WITH_RETURN	DATA_PCT	BTS_ID
1	TRAIN1	TRN_	Y	30	1
1	TEST1	TST_	N	30	2

7.1.2 SAS Macro Program

Below is the %bootstrap SAS macro that is used to generate bootstrapped training or testing datasets. Parameters for the %bootstrap macro are extracted from the metadata stored in the data dictionary for input data component.

```
%macro bootstrap(
   data=,             /* Input training dataset. */
   output = ,         /* Output for bootstrapped training dataset.
                      */
   samples=,          /* Number of resamples to generate. */
   size=,             /* Size of each resample. */
   balanced=,         /* 1 for balanced resampling; 0 for uniform
                         resampling. */
   random=            /* Seed for pseudorandom numbers. */
   );

   *** find number of observations in the input data set;
   %global _nobs;
   data _null_;
      call symput('_nobs',trim(left(put(_nobs,12.))));
      if 0 then
         set &data nobs=_nobs;
      stop;
   run;

   %if &balanced %then
      %bootsbalance(data=&data,samples=&samples,random=
      &random, output = &output);
```

```
    %else
        %bootsuni(data=&data,samples=&samples,random=&random,
        size=&size, output = &output);

%mend bootstrap;

%macro bootsbalance ( /* Balanced bootstrap resampling */
    data=,
    samples= ,
    random=,
    output =
    );

* Gleason, J.R. (1988) "Algorithms for balanced bootstrap
* simulations," American Statistician, 42, 263-266;

    data &output ;

        drop _a _cbig _ii _j _jbig _k _s;
        array _c(&_nobs) _temporary_;  /* cell counts */
        array _p(&_nobs) _temporary_;  /* pointers */
        do _j=1 to &_nobs;
            _c(_j)=&samples;
        end;
        do _j=1 to &_nobs;
            _p(_j)=_j;
        end;
        _k=&_nobs;              /* number of nonempty cells left */
        _jbig=_k;               /* index of largest cell */
        _cbig = &samples;       /* _cbig >= _c(_j) */
        do _sample_=1 to &samples;
            do _i=1 to &_nobs;
                do until(_s<=_c(_j));
                    _j=ceil(ranuni(&random)*_k);
                                /* choose a cell */
                    _s=ceil(ranuni(&random)*_cbig);
                                /* accept cell? */
                end;
                _l=_p(_j);
                _obs_=_l;
                _c(_j)+-1;
                if _j=_jbig then do;
                    _a=floor((&samples-_sample_+_k)/_k);
                    if _cbig-_c(_j)>_a then do;
                        do _ii=1 to _k;
                            if _c(_ii)>_c(_jbig) then _jbig=_ii;
                        end;
                        _cbig=_c(_jbig);
                    end;
                end;
            end;
```

```
            if _c(_j)=0 then do;
                if _jbig=_k then _jbig=_j;
                _p(_j)=_p(_k);
                _c(_j)=_c(_k);
                _k+-1;
            end;
              set &data point = _l ;
              output;
          end;
      end;
      stop;
   run;

%mend bootsbalance;
%macro bootsuni ( /* Uniform bootstrap resampling */
   data= ,
   samples= ,
   random= ,
   size= ,
   output =
   );

   %if %bquote(&size)= %then %let size=&_nobs;

   %local sample;
   %do sample=1 %to &samples;

      data _tmp_&sample ;
          sample = &sample ;
          do _i=1 to &size;
              _p=ceil(ranuni(%eval(&random+&sample))*&_nobs);
              set &data point=_p;
              output ;
          end;
          stop;
      run;
      %if &sample > 1 %then %do ;
          proc append base = _tmp_1 data = _tmp_&sample ;
          run ;
      %end ;

   %end;
   data &output ;
     set _tmp_1 ;
   run ;

   proc datasets library = work memtype=data noprint ;
      delete
      %do sample=1 %to &samples;
        _tmp_&sample
```

```
    %end ;;
  run ;
  quit ;
%mend bootsuni;
```

7.1.3 R Program

A major component of bootstrapping is being able to resample a given dataset, and in R, the function which does this is the sample function.

```
sample (data, size, replace, prob)
```

The first argument, data, is a vector containing the dataset to be resampled or the indices of the data to be resampled. The size option specifies the sample size with the default being the size of the population being resampled. The replace option determines whether the sample will be drawn with or without replacement where the default value is FALSE, i.e., without replacement. The prob option takes a vector of length equal to the dataset given in the first argument containing the probability of selection for each element of data. The default value is for a random sample where each element has an equal probability of being sampled. You can create bootstrapped samples using the sample and replicate functions:

```
Replicate (number_samples, sample())
```

7.2 Summary

This chapter described a data dictionary for input data component of AI framework. It supports code-free design and facilitates robust implementation. This chapter described tables comprising the data dictionary as well as showed the relationships among those tables. These relationships outline the process of data cleansing, outliers detection and treatment, and generation of training or testing datasets. This data dictionary is flexible enough to be extended by the reader. In the next chapters, we describe data dictionaries for design of experiments and support vector machine classification.

8

Design of Experiment for Machine Learning Component

8.1 Data Dictionary

In order to find optimal hyperparameters for supervised machine learning, multiple experiments should be conducted by data scientists. These experiments may answer the questions which class of kernel transformation to use, what features (data variables) to include, what values for hyperparameters to use, etc. Without a standardized way of managing the results of multiple experiments with machine learning method applied with different kernels, features, and hyperparameters, it will be virtually impossible to reproduce or compare the results of experiments. To achieve comparability and reproducibility of experiments the definitions of such experiments should be stored as metadata in a specially structured data (Kolosova and Berestizhevsky, 1995).

A data dictionary for the design of experiment for machine learning component is intended to define what experiments should be performed to find a combination of features set (set of data variables) and optimal hyperparameters for a specified machine learning method. This data dictionary includes, at a minimum, machine learning method name, kernel or activation function, and list of hyperparameters as well as definitions of the statistically designed experiment such as the type of experiment, statistical model, and metrics to define the quality of results of the execution of a machine learning method. The data dictionary also includes definitions of types and values of hyperparameters. There are seven tables that comprise the design of experiments for machine learning component data dictionary. They are Experiment, Metrics, Features, ML_Method, Hyperparameters_Domain, Results, and Results_Metrics. Relations among the tables of the data dictionary are presented in Figure 8.1.

The two most important reasons to define experiments through metadata are to ensure comparability across multiple experiments with different values of hyperparameters, different features sets, kernel transformations, and types of experiments as well as to ensure reproducibility of the results of experiments.

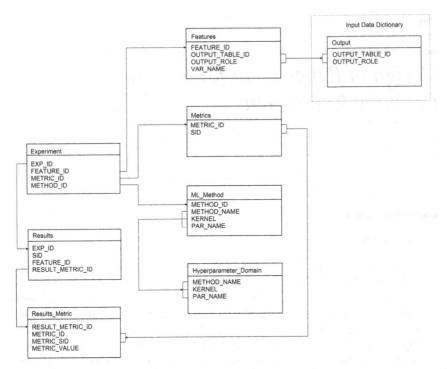

FIGURE 8.1
Data dictionary for design of experiments for machine learning component.

8.1.1　Experiment Table

The columns of the Experiment table are defined as shown in Table 8.1.

The EXP_ID column comprises the primary key of the Experiment table. The Experiment table defines what datasets should be used, with which features set, what machine learning method should be applied, and what metrics will be used to determine the quality of results of application of machine learning method. References to datasets, features sets, methods of machine learning, and quality metrics are defined in the FEATURE_ID, METHOD_ID, and METRIC_ID columns. The metadata for these references is located in the Output table that is defined in input component data dictionary (Part II, Chapter 7). The other tables will be defined below.

8.1.2　Features Table

The columns of the Features table are defined as shown in Table 8.2.

The FEATURE_ID, OUTPUT_TABLE_ID, OUTPUT_ROLE, and VAR_NAME columns comprise the primary key of the Features table. The Features table defines what variables from dataset should be used for the specified machine learning method in the Experiment table.

TABLE 8.1

Columns of the Experiment Table

Column Name	Type	Length	Description
EXP_ID	Character	8	Experiment ID
FEATURE_ID	Character	8	Reference to the primary key of the Features table
METHOD_ID	Character	8	Reference to the primary key of the ML_Method table
METRIC_ID	Character	8	Reference to the primary key of the Metrics table
DESIGN	Character	4	Type of design of experiment: FULL; FRAC
MODEL	Character	8	Type of model of possible effects of factors on response, where response is a quality metric defined in the Metrics table: MAIN—effects only, D2—main effects and second-degree interactions, D3—main effects and up to third-degree interactions

TABLE 8.2

Columns of the Features Table

Column Name	Type	Length	Description
FEATURE_ID	Character	8	Features set ID
OUTPUT_TABLE_ID	Character	8	Reference to the primary key of the Output table of input component data dictionary
OUTPUT_ROLE	Character	80	Reference to the primary key of the Output table
VAR _NAME	Character	1	Variable name from the dataset defined in the Output table with OUTPUT_TABLE_ID

8.1.3 Metrics Table

The columns of the Metrics table are defined as shown in Table 8.3.

The METRIC_ID and SID columns comprise the primary key of the Metrics table. The Metrics table defines what metrics, such as sensitivity, specificity, along with their cutoff minimum and maximum values should be used

TABLE 8.3

Columns of the Metrics Table

Column Name	Type	Length	Description
METRIC_ID	Character	8	Metric ID
SID	Character	8	Sequential number
METRIC	Character	8	Metric name
MIN_CUTOFF	Numeric	8	Minimum cutoff value for a metric below which the quality is unacceptable
MAX_CUTOFF	Numeric	8	Maximum cutoff value for a metric

to assess the quality of results of application of machine learning methods defined in the ML_Method table.

8.1.4 ML_Method Table

The columns of the ML_Method table are defined as shown in Table 8.4.

The METHOD_ID, METHOD_NAME, KERNEL, and PAR_NAME columns comprise the primary key of the ML_Method table. The ML_METHOD table defines the name of the machine learning method, its kernel transformation and references to hyperparameters associated with method and kernel transformation and their minimum and maximum values and number of groups (LEVELS column) into which the values of hyperparameters should be aggregated. This aggregation will allow specifying the factorial experiment where hyperparameters serve as factors and their discrete groups serve as levels of those factors. These hyperparameters are defined in the Hyperparameters_Domain table below.

8.1.5 Hyperparameters_Domain Table

The columns of the Hyperparameters_Domain table are defined as shown in Table 8.5.

The METHOD_NAME, KERNEL, and PAR_NAME columns comprise the primary key of the Hyperparameters_Domain table.

8.1.6 Results Table

The columns of the Results table are defined as shown in Table 8.6.

The EXP_ID and SID columns comprise the primary key of the Results table. This table contains results of experiment in terms of ML method, kernel, features, and hyperparameters and their optimal values that are located in a specified interval (level). The content of this table is produced by an SAS

TABLE 8.4

Columns of the ML_Method Table

Column Name	Type	Length	Description
METHOD_ID	Character	8	Machine learning method ID
METHOD_NAME	Character	8	Machine learning method name
KERNEL	Character	8	Kernel transformation name
PAR_NAME	Character	8	Reference to PAR_NAME column of the Hyperparameters_Domain table
MIN_VALUE	Numeric	8	Minimum value of hyperparameter
MAX_VALUE	Numeric	8	Maximum value of hyperparameter
LEVELS	Numeric	8	Number of levels (aggregation groups) for values of hyperparameters

TABLE 8.5

Columns of the Hyperparameters_Domain Table

Column name	Type	Length	Description
METHOD_NAME	Character	8	Hyperparameter ID
KERNEL	Character	8	Kernel name
PAR_NAME	Character	8	Hyperparameter name
VALUE_TYPE	Character	1	Value type: I(nteger), C(continues)
UPPER_VALUE	Numeric	8	Highest possible value of the hyperparameter
LOWER_VALUE	Numeric	8	Lowest possible value of the hyperparameter

TABLE 8.6

Columns of the Results Table

Column Name	Type	Length	Description
EXP_ID	Character	8	Experiment ID
SID	Integer	8	Sequential number ID
FEATUIRE_ID	Character	8	Feature ID
RESULT_METRIC_ID	Character	8	Result metric ID
METHOD_NAME	Character	8	ML method
KERNEL	Character	8	Kernel name
PAR_NAME	Character	8	Hyperparameter name
LEVEL_VALUE	Numeric	8	Optimal level for hyperparameter

and/or R computer program that executes statistically designed experiments defined as metadata in the five tables described above.

8.1.7 Results_Metrics Table

The columns of the Results_Metrics table are defined as shown in Table 8.7.
The RESULTS_METRIC_ID, METRIC_ID, and METRIC_SID comprise the primary key of the Results_Metrics table. This table contains values of quality metrics of classification results produced during statistically designed experiments intended to find optimal values (levels) of hyperparameters. The content of this table is produced by an SAS and/or R computer program

TABLE 8.7

Columns of the Results_Metrics Table

Column Name	Type	Length	Description
RESULT_METRIC_ID	Character	8	Result metric ID
METRIC ID	Character	8	Metric ID
METRIC_SID	Integer	8	Sequential number of the Metrics table
METRIC_value	Numeric	8	Result metric value

that executes statistically designed experiments defined as metadata in the five tables described above.

To get a better understanding of what all the above-described metadata definitions mean, we created an example of design of experiments for machine learning data dictionary table pre-filled with metadata.

Tables 8.8–8.14 show the examples of how the content of the data dictionary tables for design of experiments for machine learning component may look like.

TABLE 8.8

Content of the Experiment Table

EXP_ID	FEATURE_ID	METHOD_ID	METRIC_ID	DESIGN	MODEL
001	001	001	001	FULL	MAIN
002	001	002	002	FRAC	D2

TABLE 8.9

Content of the Features Table

FEATURE_ID	OUTPUT_TABLE_ID	OUTPUT_ROLE	VAR_NAME
001	1	TRAIN1	VarOne
001	1	TRAIN1	VarTwo
001	1	TRAIN1	VarThree

TABLE 8.10

Content of the Metrics Table

METRIC_ID	SID	METRIC	MIN_CUTOFF	MAX_CUTOFF
001	1	TPR	0.8	0.97
001	2	TNR	0.75	0.96
002	1	TPR	0.79	0.89

TABLE 8.11

Content of the ML_Method Table

METHOD_ID	METHOD_NAME	KERNEL	PAR_NAME	MIN_VALUE	MAX_VALUE	LEVELS
001	SVM	RADIAL	GAMMA	0.05	10	3
001	SVM	RADIAL	COST	0.05	0.09	2

TABLE 8.12

Content of the Hyperparameters_Domain Table

METHOD_NAME	KERNEL	PAR_ NAME	VALUE_ TYPE	UPPPER_ VALUE	LOWER_ VALUE
SVM	POLYNOM	DEGREE	I	5	2
SVM	POLYNOM	COST	C	100,000	0
SVM	RADIAL	GAMMA	C	10	0.05
SVM	RADIAL	COST	C	100,000	0

TABLE 8.13

Content of the Results Table

EXP_ID	SID	FEATURE_ID	RESULT_ METRIC_ID	METHOD_ NAME	KERNEL	PAR_ NAME	LEVEL_ VALUE
001	1	001	1	SVM	RADIAL	GAMMA	1
001	2	001	1	SVM	RADIAL	COST	3

TABLE 8.14

Content of the Results_Metrics Table

RESULTS_METRIC_ID	METRIC_ID	METRIC_SID	METRIC_VALUE
1	001	1	0.87
1	001	2	0.79
2	002	1	0.92

8.2 SAS Macro Program

The source code of the SAS macro programs can be continuously improved and extended after the book has been published. It is located on GitHub:
https://github.com/smlof/Supervised-Machine-Learning--Optimization-Framework

8.3 R Programs

The source code of the R programs can be continuously improved and extended after the book has been published. It is located on GitHub:
https://github.com/smlof/Supervised-Machine-Learning--Optimization-Framework

8.4 Summary

This chapter described a data dictionary for design of experiments for machine learning component of the AI framework. The main goal of this data dictionary is to ensure reproducibility and comparability of computer-based experiments that result in delivering maximum values of quality metrics of results of application of machine learning method conditionally on values for hyperparameters for the specified machine learning method. These values of hyperparameters can be defined as optimal. This chapter described tables comprising the data dictionary as well as showed relationships among the tables. These relationships outline the process of selection of features set, creating design of experiment using hyperparameters as factors with discrete levels (groups of aggregated values of these parameters from minimum to maximum values they can get), metrics that can be used to assess the quality of results of machine learning method applied to a specified features set and combination of levels of hyperparameters. This data dictionary is flexible enough to be extended by the reader.

Reference

Kolosova, T., and Berestizhevsky, S. 1995. *Table-Driven Strategies for Rapid SAS Applications Development.* Cary, NC: SAS Institute, Inc.

9

"Contaminated" Training Datasets Component

Support vector machine (SVM) is a highly developed classification method that is widely used in real-world applications; however, SVM has drawbacks. The important feature of the SVM is that the separating hyperplane is determined mainly from misclassified cases. The last means that the most misclassified cases significantly affect the quality of the classifier, and meaning of that is the SVM with optimal hyperparameters identified on "clean" training datasets is extremely fragile to the presence of outliers; for example, some of the labels of training datasets are corrupted. In this chapter, we describe an approach to test the robustness of the classifier and identify the fixed proportion of data with "label contamination" ("label flipping" or "label corruption") in training datasets, in which the classifier with newly identified optimal hyperparameters still provides useful classification results. It appears that "label contamination" of training datasets helps to create the classifier that performs sufficiently well when it classifies unseen data.

In order to find optimal hyperparameters for supervised machine learning in the case of a fixed proportion of "label contamination," statistically designed experiments should be defined and conducted. That said, the research of theoretical methods for identifying an optimal fixed proportion of "label contamination" that would not impact severely the quality of classifiers would be an interesting issue for future work.

9.1 Data Dictionary

In order to find optimal hyperparameters for classifiers trained on "contaminated" datasets, multiple experiments should be conducted by data scientists. These experiments may answer the questions which class of kernel transformation to use, what features (data variables) to include, what values for hyperparameters to use, etc. Without a standardized way of managing the results of multiple experiments with the machine learning method applied with different kernels, features, and hyperparameters, it will be virtually impossible to reproduce or compare results of experiments. To achieve comparability and reproducibility of experiments, the definitions

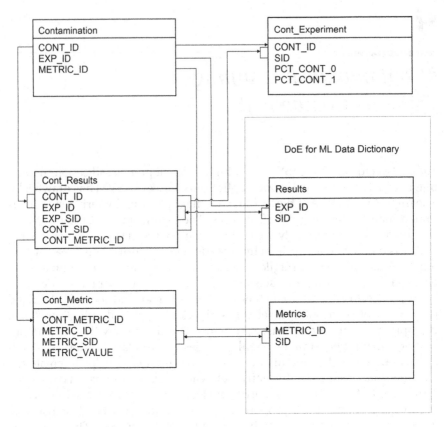

FIGURE 9.1
Data Dictionary for Contaminated Training Datasets component.

of such experiments should be stored as metadata in a specially structured data dictionary (Kolosova and Berestizhevsky, 1995).

There are four tables that comprise the Contaminated Training Datasets component data dictionary. They are Contamination, Cont_Experiment, Cont_Results, and Cont_Metric. Relations among tables of the data dictionary are presented in Figure 9.1.

9.1.1 Contamination Table

The columns of the Contamination table are defined as shown in Table 9.1.

The CONT_ID and EXP_ID columns comprise the primary key of the Contamination table. This table contains metadata that defines which experiment should be performed for a specified "contamination" of training dataset and which metrics should be calculated to measure the quality of classification results.

TABLE 9.1

Columns of the Contamination Table

Column Name	Type	Length	Description
CONT_ID	Character	8	Contamination ID
EXP_ID	Character	8	Reference to primary key of Results table of Data Dictionary for Design of Experiment for Machine Learning
METRIC_ID	Character	8	Reference to primary key of Metric table of Data Dictionary for Design of Experiment for Machine Learning

9.1.2 Cont_Experiment Table

The columns of the Cont_Experiment table are defined as shown in Table 9.2.

The CONT_ID and SID columns comprise the primary key of the Cont_Results table. This table contains metadata that defines which "contamination" of labels of training datasets should be performed.

9.1.3 Cont_Results Table

The columns of the Cont_Results table are defined as shown in Table 9.3.

TABLE 9.2

Columns of the Cont_Experiment Table

Column Name	Type	Length	Description
CONT_ID	Character	8	Contamination ID
SID	Integer	8	Sequential number
PCT_CONT_0	Integer	8	Percent of contamination of labels for 0
PCT_CONT_1	Integer	8	Percent of contamination of labels for 1

TABLE 9.3

Columns of the Cont_Results Table

Column name	Type	Length	Description
CONT_ID	Character	8	Contamination ID
EXP_ID	Character	8	Reference to the primary key of Results table of Data Dictionary for Design of Experiment for Machine Learning
EXP_SID	Character	8	Reference to the primary key of Results table of Data Dictionary for Design of Experiment for Machine Learning
CONT_SID	Character	8	Sequential number of contamination
CONT_METRIC_ID	Character	4	Contamination Metric ID

TABLE 9.4

Columns of the Cont_Metric Table

Column Name	Type	Length	Description
CONT_METRIC_ID	Character	8	Contamination Metric ID
METRIC_ID	Character	8	Reference to primary key of Metric table of Data Dictionary for Design of Experiment for Machine Learning
METRIC_SID	Character	8	Sequential number for METRIC_ID
METRIC_VALUE	Integer	8	Value of specified metric

The CONT_ID, EXP_ID, EXP_SID, and CONT_SID columns comprise the primary key of the Cont_Results table. The content of this table is produced by a process that executes a statistically designed experiment applied to contaminated training datasets.

9.1.4 Cont_Metric Table

The columns of the Cont_Metric table are defined as shown in Table 9.4.

The CONT_METRIC_ID, METRIC_ID, and METRIC_SID columns comprise the primary key of the Cont_Metric table. The content of this table is produced by a process that executes a statistically designed experiment using contaminated training datasets.

9.2 SAS Macro Program

The source code of the SAS macro programs can be continuously improved and extended after the book has been published. It is located on GitHub:

https://github.com/smlof/Supervised-Machine-Learning--Optimization-Framework

9.3 R Programs

The source code of the R programs can be continuously improved and extended after the book has been published. It is located on GitHub:

https://github.com/smlof/Supervised-Machine-Learning--Optimization-Framework

9.4 Summary

This chapter described a data dictionary for the Contaminated Training Datasets component of the AI framework. The main goal of this data dictionary is to ensure reproducibility and comparability of computer-based experiments. These experiments intended to find conditions (hyperparameter value and feature set) that deliver maximum possible values of quality metrics of classification. Also, this data dictionary helps to identify percentages of "contamination" of labels of training datasets for which classifiers still produce a maximum possible quality of classification. Eventually, knowing the hyperparameters of classifiers that produce quality results on "contaminated" training datasets helps to ensure quality results of classification for unseen data. The last is the ultimate goal of machine learning classification.

Reference

Kolosova, T., and Berestizhevsky, S. 1995. *Table-Driven Strategies for Rapid SAS Applications Development.* Cary, NC: SAS Institute, Inc.

Part III

Part III

10

Insurance Industry: Underwriters' Decision-Making Process

10.1 Introduction

Traditionally, underwriting decisions were considered the result of the judgment, experience, and "gut feelings." Apprenticeship was a customary way for a new or young underwriter to gain the necessary experience. However, it took a long time for senior underwriters to pass their knowledge and expertise to a younger generation, who would then do the same for the next generation.

One approach to improve underwriting training was to analyze in detail the steps in the underwriting decision-making process and creating guidelines (Chidambaran et al., 1997). Thinking of underwriting as an application of the specific steps in the decision-making process clarified its nature but formed a rigid structure that lost the flexibility of experienced underwriters. Thus, the approach fell short of expectations. The industry still struggles with the pace of knowledge and experience delivery to the next generation of underwriters.

Also, insurers began to learn that experience is not necessarily the best or most efficient teacher—it is time-consuming and expensive, and subjective. Bad habits, as well as good habits, could be passed on through the apprentice system.

The improvement of the underwriting decision-making process can be made through the following three steps:

1. Identification of the best underwriters using various relevant performance metrics
2. Training AI algorithms utilizing the knowledge of the best-performing underwriters, and
3. Deployment of AI solutions for automation of the underwriting decision-making process.

10.2 Review of Underwriters' Performance

In this section, we analyze the performance of underwriters using several existing metrics (Kunreuther et al., 1995) and a novel one proposed here. We briefly explain the characteristics of these metrics and then apply them to the real-life data provided by a workers' compensation insurer.

10.2.1 Metrics of Underwriters' Performance

Different metrics can be used to evaluate underwriters' performance, for example, hit ratio, conversion rate, and time-to-deal. We also propose a novel metric that can measure underwriters' performance dynamically.

10.2.1.1 Hit Ratio

The hit ratio (HitR) metric presents an underwriter's performance that depends on the number of assigned applications. This metric is calculated over some period, which is defined depending on the workload of underwriters and expected time to close the deal (bind an insurance application), or the expiration period for a quote.

$$\text{HitR}_{ti} = \frac{\text{BU}_{ti}}{\text{AU}_{ti}}$$

where:
  HitR_{ti}—hit ratio of the underwriter i during the period t
  BU_{ti}—number of bound applications (deals closed) by the underwriter i during the period t
  AU_{ti}—number of applications assigned to the underwriter i during the period t.

A significant disadvantage of the hit ratio metric is that it does not take into consideration the complexity of insurance applications. Very often, underwriters reject insurance applications out of hand, or after short evaluation against the insurer's rules and policies. So, hit ratio also depends on the complexity of the assigned to the underwriter insurance applications. Thus, instead of hit ratio, the conversion rate metric would be an adequate way of measuring underwriters' performance.

10.2.1.2 Conversion Rate

The conversion rate metric (CR_{ti}) presents an underwriter's performance based on only those applications that were quoted by the underwriter. In other words, quoted applications are those that were not rejected by the insurer (underwriter). This metric is calculated over a period defined by expected time to close the deal (to bound the application, or accept by a customer the proposed quote), or the expiration time for a quote:

$$CR_{ti} = \frac{BU_{ti}}{QAU_{ti}}$$

where:

CR_{ti}—the conversion rate of the underwriter i during the period t

BU_{ti}—number of bound applications for the underwriter i during the period t

QAU_{ti}—number of applications quoted by the underwriter i during the period t.

This metric is usually calculated over an extended period because underwriters would want to wait for the outcome of all applications, including those that are still in the process of negotiation with customers/prospects. Thus, this metric, though providing an informative backward-looking picture, cannot be used as an operational metric, for example, monthly. Another problem with this metric is that it is not taking into consideration the workload of the specific underwriter. As an example, the underwriter who bound 1 out of 5 quoted applications will have the same $CR_{ti} = \frac{1}{5} = 0.2$ as a person who bound 10 out of 50 quoted applications $CR_{ti} = \frac{10}{50} = 0.2$.

We propose a new operational metric that addresses these issues.

10.2.1.3 Dynamic Conversion Rate

We propose a new operational metric called dynamic conversion rate. The dynamic conversion rate measures conversion continuously, allowing for the monitoring of underwriters' performance as well as managing optimal underwriters' workload.

The dynamic conversion rate metric (DCR_{mi}) evaluates a monthly underwriter's performance based on information available in the month of interest. This metric considers an underwriter as a service provider that has incoming requests (quoted applications) and outgoing results (bound applications). In other words, each month, an underwriter has several quoted applications that are in the queue to be finalized: to be bound or rejected by a customer/prospect. Because DCR_{mi} measures performance every month, underwriters with low workload are penalized for inactive months. The dynamic conversion rate metric is calculated according to the following formula:

$$DCR_{mi} = \frac{\ln\left(BU_{mi} + 1\right)}{\ln(QQAU_{mi} + 1)}$$

where:

DCR$_{mi}$—the dynamic conversion rate of the underwriter i during the month m

BU$_{ti}$—number of bound applications for the underwriter i during the month m

QQAU$_{ti}$—number of applications quoted by the underwriter i and awaiting resolution (in the queue) during the month m.

This metric gets values between 0 and 1; DCR$_{mi}$ equals 0 when there are no bound applications in a current month and equals 1 when all quoted until now applications were bound in the current month. Thus, this metric presents an ongoing performance of underwriters, providing an operational measure that can be used for efficient management of the underwriting process. Also, the DCR$_{mi}$ metric takes into consideration the workload of an underwriter. Table 10.1 demonstrates the benefits of the DCR$_{mi}$ metric using the same example as we used above for the CR$_{ti}$ conversion rate metric.

As shown in Table 10.1, DCR$_{mi}$ is higher for an underwriter B who has the same conversion rate CR$_{ti}$ = 0.2 as underwriter A, but a larger number of quoted and bound applications.

10.2.1.4 Time-to-Deal

The time-to-deal (TTD) metric evaluates an underwriter's efficiency in the closing of the deals. Time-to-deal is calculated as a time interval in days between the date when the insurance application was assigned to an underwriter and the date when the application was bound. The TTD metric is censored by a quote expiration date. TTD can be averaged over an extended period like 6 months or a year, but in this case, it is not operational and cannot be used for successful management of underwriting process. We propose to use monthly time-to-deal metric TTD$_{mi}$, which is an average time-to-deal calculated for the underwriter i during the month m.

$$TTD_{mi} = \frac{\sum_{j=1}^{n} TTD_{mij}}{n}$$

where:

TTD$_{mi}$—an average time-to-deal of the underwriter i during the month m

TABLE 10.1

Comparison of CR$_{ti}$ and DCR$_{mi}$ Metrics

Underwriter	Quoted Applications (QQAU$_{ti}$)	Bound Applications (BU$_{ti}$)	CR$_{ti}$	DCR$_{mi}$
A	5	1	0.2	0.38685
B	50	10	0.2	0.60987

TTD_{mij}—a time-to-deal of the bound application j out of n applications bound by the underwriter i during the month m.

10.2.2 Analysis of Underwriters' Performance

Using the above-described metrics, we present the results of a study of the performance of underwriters for workers' compensation line of insurance. In this study, we use data related to new workers' compensation applications for a period of 44 months.

10.2.2.1 Data Description

Data used for the study of underwrites performance evaluation include the following:

- dates of applications submission, dates of quotes, and dates of policy binding, if any;
- parameters of insurance applications such as industry, state, and number of employees;
- underwriters to whom applications were assigned.

Before the analysis, the data was "cleansed" and prepared in the following way: (a) Applications that required more than three months to be bound were removed from the study (less than 5% of all bound applications). (b) Underwriters who were assigned less than 2% of all applications during the analyzed period of 44 months were excluded as well.

10.2.2.2 Application Flow

Underwriters' performance is evaluated conditionally on insurance application flow, and the process of assignment applications to underwriters. Thus, before estimating dynamic conversion rate and time-to-deal per underwriter, the analysis of application flow and the process of application assignment should be performed.

Distribution of an average number of applications per month demonstrated seasonality with a somewhat increased number of applications in March and October, and with a significant drop in December (Figure 10.1). Identification of this seasonality in the submission process helps create a balanced workload for underwriters.

Distribution of an average number of applications per weekday demonstrated a slightly lower number of applications on Mondays (Figure 10.2).

The analysis indicates that the application submission flow has some fluctuations among months and among weekdays which are easily manageable.

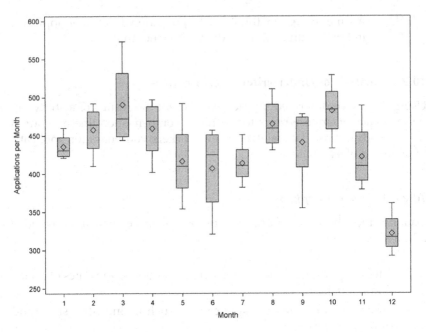

FIGURE 10.1
Average number of monthly submitted applications.

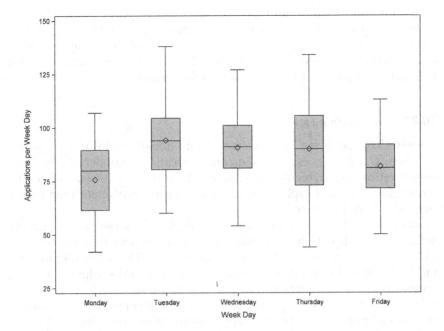

FIGURE 10.2
Average number of weekday submitted applications.

However, as shown in Figure 10.3, the workload of underwriters is far from being balanced. It makes an analysis of underwriters' performance more complicated.

On the one hand, it would be reasonable to assign more applications to underwriters with better performance. However, on the other hand, it is possible that fewer applications allow the underwriters to analyze the situation more comprehensively, to have a more in-depth negotiation with a customer/prospect—and as a result, to demonstrate better performance. We will discuss this hypothesis further.

10.2.2.3 Dynamic Conversion Rate per Underwriter

The dynamic conversion rate DCR_{mi} per underwriter was calculated over the observation period of 44 months. The best underwriters, with an average dynamic conversion rate above 0.12, comprise about 8% of all underwriters. As shown in Figure 10.4, top performers have a significantly higher dynamic conversion rate than the rest of the underwriters.

We analyzed the relations of the number of assigned applications to an underwriter with their performance measured by DCR_{mi}. For the top performers, the average number of monthly assigned applications per underwriter is 13 (median is 12). For the rest of the underwriters, the average

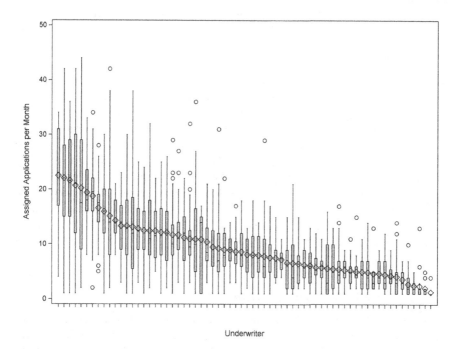

FIGURE 10.3
Average monthly application assignment per underwriter.

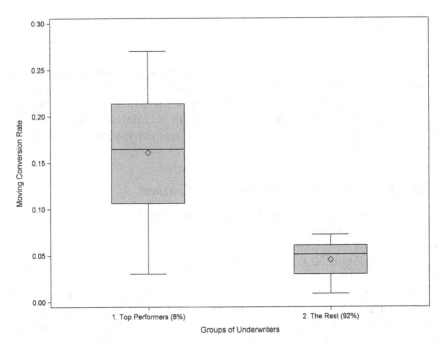

FIGURE 10.4
Comparison of dynamic conversion rates (moving conversion rates) between two groups of underwriters.

number of monthly assigned applications per underwriter is 10 (median is 8). We performed a comparison of empirical distributions of numbers of monthly assigned applications per underwriter between the group of the top performers and the group of the rest of the underwriters using the Kolmogorov–Smirnov two-sample test. The asymptotic p-value for the Kolmogorov–Smirnov test is <.0001. This indicates that we should reject the null hypothesis that the distributions of numbers of monthly assigned applications per underwriter are the same for the two groups of underwriters. This demonstrates that, indeed, top underwriters (according to the dynamic conversion rate metric) are assigned more applications than other underwriters, and regardless of increased workload, they still perform substantially better than others.

10.2.2.4 Time-to-Deal per Underwriter

The time-to-deal metric (TTD_{mi}) was calculated per each underwriter. Time-to-deal varies significantly, from 0 to 90 days (where 90 days is an expiration term of a quote), with an average of 32 days and a median of 29 days. To compare time-to-deal of top performers vs. the rest of the underwriters, we performed a comparison of empirical distributions of TTD_{mi} between these two groups using the Kolmogorov–Smirnov two-sample test. The asymptotic p-value for the Kolmogorov–Smirnov test is 0.1255. This indicates that we cannot reject the null hypothesis that the distributions are the same for the two groups of underwriters. Both TTD_{mi} and DCR_{mi} can be used as operational metrics for ongoing monitoring and management of underwriters' performance.

10.3 Traditional Approach to Knowledge Delivery

It is an ongoing effort for insurers to improve the quality and consistency of the underwriting process. For years, sound underwriting decisions were considered the result of the judgment, experience, and "gut feelings." The only way for a new or young underwriter to gain "judgment" was by working under the guidance of a successful senior underwriter and gradually gaining experience (Randall, 1994).

Traditional ways to validate performance include underwriting file reviews, or audits, in an attempt to improve quality and consistency. These audits attempt to determine the effectiveness of underwriters. The audit process usually consists of six steps, validating each phase in the underwriting decision process:

1. Gather data on an applicant's loss exposures and associated hazards.
2. Evaluate the exposures and hazards.
3. Develop and evaluate underwriting options.
4. Select an underwriting option, including selecting the appropriate premium.
5. Implement the underwriting decision.
6. Monitor the loss exposures.

This audit may identify deficiencies in underwriting processes and provide a basis for additional coaching, improved documentation, and enhanced quality reviews. However, each step in this process is validated subjectively and qualitatively and is based on two assumptions. The first one is appropriate guidelines, which are usually updated infrequently and become obsolete with time. The second assumption is the existence of a matured management process, where the assignment of applications to underwriters is balanced and identification of a problem and intervention to fix it are timely.

Unfortunately, in today's dynamic, large-scale, and innovative environment, these assumptions are often not valid. Also, the growing need for underwriters requires that training is streamlined, consistent, and fast. Though thinking of underwriting as steps in the decision-making process clarified its nature, it only led to static guidelines that did not include experience, sound judgment, and intuition of the top underwriters.

From our analysis of underwriters' performance, it is evident that a small number of underwriters perform significantly better than others, despite the guidelines and audits. More experienced and knowledgeable underwriters apply skills that cannot be captured by guidelines, and their expert knowledge allows them to intuitively analyze multiple factors and their combinations to produce a better decision.

Artificial intelligence approaches based on machine learning algorithms can be used to model the decision-making process of the best underwriters, and then, these models can be applied to facilitate and automate the underwriting decision-making process.

10.4 Anatomy of Artificial Intelligence Solution

The underwriting decision-making process includes the following components:

- Underwriters, who are assigned to work on insurance applications (in our example, workers' compensation line of insurance)
- Insurance applications submitted by various customers/prospects along with a description of different characteristics of the business, including business size and turnover
- Insurance products offered by insurer for workers' compensation along with their conditions, premium size, limitations, etc.
- Customer/prospect responses to underwriters' quotes: accepting or rejecting the proposed insurance offer

To formalize the underwriting decision-making process, we start with creating the appropriate data structure.

10.4.1 Data Structure

Let $i = 1...n$ denote underwriters, $j = 1...m$ denote applications, and $k = 1...p$ denote insurance products for workers' compensation. Each application is described by a set of features (parameters), for example, the number of employees in the company, ages of the employees, the industry code, the state/province, and number of claims in the previous year. These features are standardized and presented by a vector $A_j = \{a_{j1}, a_{j2}, ... a_{jM}\}$, where M is the number of insurance application features (parameters or characteristics).

Each insurance product is described by a set of features (parameters or characteristics), for example, the conditions covered by the policy, the deductible amount, and premium amount. Insurance product features are presented by a vector $B_k = \{b_{k1}, b_{k2}, \ldots b_{kP}\}$, where P is the number of insurance product features. These features are standardized for all products, and if, for example, feature b_{k2} is not presented in the product k, then b_{k2} is assigned 0.

Let Y_{ijk} denote an outcome of each quoted insurance application—1 means bound, and 0 means rejected. This structure allows capturing multiple quotes made by the same underwriter as a response to the same insurance application. For example, the above-described data structure may look like that in Table 10.2.

10.4.2 Classification Approach

In the broadest sense, we try to solve the problem of predicting whether or not the customer/prospect will accept the insurance product offered by an insurer. We need to find a fit between the features (parameters or characteristics) of the insurance application and the features of the insurance product and its price. The substantial challenge is that data is very unbalanced: in our study, bound insurance applications comprise about 9% of the quoted applications.

We approach the problem using the support vector machine (SVM)—a proven classification method that is also well suited to our problem because it can directly minimize the classification error without requiring a statistical model (Cortes and Vapnik, 1995). SVM can handle the massive feature space, and it is guaranteed to find a globally optimal separating hyperplane if it exists. Also, SVM models are built on the boundary cases, which allow them to handle missing data. SVM is described in Chapter 2 of Part I.

10.4.3 Bias–Variance Trade-Off and SVM Hyperparameters

The bias is an error caused by erroneous assumptions in the learning algorithm. High bias means an underfitting between features and outputs. The variance is an error produced from variations in the sensitivity metric in response to small fluctuations in the training sets. High variance caused overfitting between features and outputs.

The bias–variance trade-off is a central problem in supervised learning. Unfortunately, it is impossible to create a classification model that both accurately captures the regularities in its training data and generalizes well to unseen data. In statistics and in machine learning, the bias–variance trade-off is the property of a set of predictive models, whereby models with lower bias in parameter estimation have a higher variance of the parameter estimates across samples and vice versa. The bias–variance problem is the

TABLE 10.2

Example of Data Structure

Underwriter i	Application j	Features of Application					Product k	Features of Product			Outcome Y_{ijk}
		Number of Employees a_{j1}	% of Employees Under 21 a_{j2}	% of Employees Over 50 a_{j3}	...	Claims in the Last Year a_{jM}		Deductible Amount b_{k1}	...	Premium Amount b_{kP}	
...	...	...	...	...	...	...	...	...	...	...	...
10	100	30	5	3	...	0	1001	10,000	...	1,000	1
10	105	100	2	10	...	5	1001	40,000	...	1,500	0
10	105	100	2	10	...	5	1003	30,000	...	1,150	1
...	...	...	...	...	...	...	...	...	...	...	...
13	201	70	6	11	...	3	1002	22,000	...	1,500	0
13	202	47	4	9	...	1	1002	11,000	...	900	0
...	...	...	...	...	...	...	...	...	...	...	...

conflict in trying to simultaneously minimize these two sources of error that prevent supervised learning algorithms like SVM from generalizing beyond their training sets.

As described in Chapter 2 of Part I, large C leads to low bias and high variance, while small C leads to high bias and low variance. Large d or γ leads to high bias and small variance, while small d or γ leads to low bias and high variance. For example, if γ is large (variance is small), it means that the SVM results do not have a wide influence and cannot be generalized beyond its training sets.

Because SVM does not model the data distribution, but instead directly minimizes the classification error, the output produced by SVM is a binary decision. Although a binary decision is sufficient for many problems, it is difficult to arrive at a meaningful bias–variance trade-off. This concern can be alleviated by applying penalized logistic regression to SVM output (see Chapter 2 of Part I). The benefit of using penalized logistic regression to SVM output is making results more generalizable (e.g., decrease overfitting). In classification, SVM, along with penalized logistic regression, creates output probabilities of class membership rather than point estimates produced by SVM alone. Rephrasing a classification problem as a probabilistic classification creates conditions for application of bias–variance decomposition that was originally formulated for least-squares regression.

10.4.4 Building the Classifier

Our proposed AI solution attempts to "learn" from the best underwriters. Thus in our study, we used only data related to the top performers comprising 8% of all underwriters. The data we worked with was structured to express all problems of similar nature by broadly including all features (characteristics or parameters) of the insurance application and the offered product and price, along with binary values Y (equivalent to acceptance or rejection of a quote by a prospect) as presented in Table 10.2.

To avoid bias in our analysis, we created training and test datasets from the original dataset using the "m-out-of-n" bootstrap method with replacement (Bickel and Ren, 1995). This method replaced complicated and often inaccurate approximations of biases, variances, and other measures of uncertainty by computer simulations. In simple situations, the uncertainty of an estimate may be gauged by analytical calculation based on an assumed probability model for the available data. But in more complicated problems, this approach can be tedious and difficult, and its results are potentially misleading if inappropriate assumptions or simplifications have been made. Applying the "m-out-of-n" bootstrap method with replacement, we select 2,000 training datasets that have the same distribution of outcome (reject/accept of insurance quote) as the original dataset of the top performers' data. We also randomly select test datasets.

Step 1

As a first step, we create a statistically designed full-factorial experiment to find out optimal hyperparameters of SVM: kernel transformation function (radial, polynomial, and sigmoid kernel types), the kernel hyperparameter (d or γ), and the algorithm hyperparameter—C. For continues values of C, and d or γ hyperparameter, we aggregate their values in multiple groups. The FACTEX procedure can be used to create a full-factorial experiment. For example, if we decide to create six groups of values for C and γ hyperparameters, our full-factorial design consists of $3^1 * 6^2 = 108$ combinations (runs).

Step 2

For each of the combinations presented in our full-factorial design, we execute the HPSVM procedure of SAS as follows:

```
%macro SVM (data=training_set, c=100, gamma=2, kernel = rbf,
input_app = num_emp pct_under_21 pct_over_50 claims, input_
prod = deduct premium output = output, target=outcome);
   PROC HPSVM DATA = &data
      C = &c /* the penalty value */
      MAXITER=10000 /* the maximum number of iterations */
      INPUT &input_app &input_prod / LEVEL=INTERVAL;
      KERNEL &kernel
      %if %trim(%left(&kernel)) = rbf or
         %trim(%left(&kernel)) = sigmoid %then
         %do;
            / k_par = &gamma;
         %end;
      OUTPUT &output;
      TARGET &target;
   run;
%mend SVM;
```

As the final classifier, we select one that maximized sensitivity (the proportion of bound applications that were correctly classified) and specificity (the proportion of rejected quotes that were correctly classified) when applied to the training datasets.

Step 3

As a next step, to add a probabilistic element to the SVM results, we apply Firth's penalized logistic regression to the outcome of the SVM algorithm (Firth, 1993). SVM produces complete separation data. Firth's penalized likelihood approach is a method of addressing issues of separability.

Firth's method is available in the LOGISTIC procedure and only for binary logistic models. It replaces the usual score equation:

$$g(\beta_j) = \sum_{i=1}^{n} (y_i - \pi_i) x_{ij} = 0 \quad (j = 1, \ldots, p)$$

where p is the number of parameters in the model, with the modified score equation:

$$g(\beta_j) = \sum_{i=1}^{n} (y_i - \pi_i + h_i (0.5 - \pi_i)) x_{ij} = 0 \quad (j = 1, \ldots, p)$$

where h_i are the ith diagonal elements of the hat matrix:

$$W^{\frac{1}{2}} X (X'WX)^{-1} X'W^{\frac{1}{2}}$$

$$W = \text{diag}\{\pi_i (1 - \pi_i)\}$$

The SAS code below implements logistic regression with Firth's correction for bias:

```
PROC LOGISTIC data = SVM_Decision ;
    CLASS Outcome ;
    MODEL Outcome (event='1') = Decision_Function / CL FIRTH ;
RUN ;
```

Since the results of logistic regression are probabilities, we chose a cutoff or a threshold value to classify cases. The probabilities above this threshold value are classified as 1 and those below as 0. For the probabilities generated by the Firth penalized logistic regression, we search for such a threshold value that it further increases the sensitivity and specificity of the results produced by SVM.

We perform the analysis for each of the 2,000 training datasets. As the final classifier, we select an SVM classifier with adjustment based on the Firth penalized logistic regression that, on average, produces higher sensitivity and specificity metrics.

After applying different kernel types and various SVM hyperparameters (according to the full-factorial design of experiment described above), the best results occur with a radial kernel having a cost parameter $C \in [1000; 3000]$ and $\gamma \in [0.2; 0.6]$. The sensitivity of the classification results equals 0.987 on average, and the specificity equals 0.799 on average.

Then, we identify a threshold that improved classification based on the results of the Firth penalized logistic regression. The threshold value of 0.76

used to classify the results of the Firth regression improved the specificity to the level of 0.847 on average at the cost of a minor loss in sensitivity—which became equal to 0.973 on average.

10.4.5 "Contamination" of Training Datasets

Now, we address the variance in the sensitivity metric caused by small fluctuations in the training datasets. To avoid high variance that caused overfitting between features and outputs, we perform "contamination" of training datasets. The "contamination" is performed as changes in small percent of labels of training datasets, planned according to the design of mixture experiments. In mixture experiments, we vary the proportions, as in our example—proportions of different outcomes. The details of the design of mixture experiments are out of the scope of this book, and we only state that this approach allows performing controlled "contamination" of data. After the training dataset is contaminated according to the design, we perform again Steps 1–3 where radial kernel having a cost parameter $C \in [1000; 3000]$ and $\gamma \in [0.2; 0.6]$ is considered as a starting point of the process. As a result, a robust classifier is built.

Applying the developed classifier to test datasets, we received the following results: specificity equals 0.859, and sensitivity equals 0.851.

10.4.6 Experimental Results

To test the efficiency of the solution, we used the following historical data:

1. properties/features of insurance products and prices that were offered in regard to insurance applications;
2. properties/features of insurance applications to which insurance products were offered;
3. data about acceptance or rejection of insurance products by clients.

The data for the experiment were selected in such a way that all offers were rejected by customers/prospects. In other words, the conversion rate in this data sample was 0%.

The objective of the experiment was to apply the developed AI solution to find what insurance products and prices would fit the insurance applications in such a way that they would be accepted by customers/prospects. Each insurance application in the sample data received one quote for an insurance product, in rare cases two quotes. In our experiment, however, each application was combined with all possible insurance products and their prices. An example of such data is presented in Table 10.3, where:

TABLE 10.3

Example of Historical Application Data with Different Product Offers

Application j	Features of Application					Product k	Features of Product			Classification Outcome Y_{ijk}
	Number of Employees a_{j1}	% of Employees Under 21 a_{j2}	% of Employees Over 50 a_{j3}	...	Claims in the Last Year a_{jM}		Deductible Amount b_{k1}	...	Premium Amount b_{kP}	
...	...	...	...	...	...	...	...	...	...	
201	70	6	11	...	3	1002	22,000	...	1,500	0
201	70	6	11	...	3	1001	20,000	...	1,250	0
201	70	6	11	...	3	1003	25,000	...	1,150	1
201	70	6	11	...	3	1004	28,000	...	1,000	0
202	47	4	9	...	1	1002	11,000	...	900	0
202	47	4	9	...	1	1001	10,000	...	900	0
202	47	4	9	...	1	1003	13,000	...	850	1
202	47	4	9	...	1	1004	15,000	...	800	1
...	...	...	...	...	...	...	...	...	T...	

- gray-shadowed rows present insurance applications and associated insurance products that were rejected by the customers, and
- white rows present all other possible products that could be offered.

This data was used as an input to our AI solution. The column entitled "Classification outcome *Yijk*" of Table 10.3 contains the results of applied AI solution. Value 1 in this column identifies products that with high probability would be accepted by the customer. Relevant quotes for insurance products and prices identified by AI solution are expected to lead to the conversion rate of 14.7%. In other words, the experiment identified the applications and the products that would most likely lead to binding of the insurance applications.

10.5 Summary

The insurance industry still struggles with the pace of knowledge delivery from experienced underwriters to young ones. The disadvantage of the currently used methods, based on apprenticeship, is the subjectivity of underwriters' decisions. Measurement of underwriters' performance also suffers from lack of efficiency, as existing metrics are delayed and not operational. We analyzed underwriters' performance metrics and proposed a novel one, dynamic conversion rate, which allows measuring and monitoring the operational performance of underwriters (weekly, monthly, etc.)

This chapter described the proposed AI solution intended to improve the underwriting decision-making process. The solution is based on the SVM machine learning method and its improvements. Overall, with the reasonably efficient classifier built by SVM, and improved with a threshold of Firth's penalized logistic regression, satisfactory results can be achieved. The chapter showed that if our AI solution would be used in quoting insurance applications (in our study, workers' compensation insurance), then it could yield incremental improvement in application binding by 14.7%.

References

Bickel, P. J., and Ren, J. J. 1995. The 'm Out of n' bootstrap and goodness of fit tests with double censored data. In *Robust Statistics, Data Analysis and Computer Intensive Methods*, 35–47. New York, NY: Springer.

Chidambaran, N. K., Pugel, T. A., and Saunders, A. 1997. An investigation of the performance of the U.S. property-liability insurance industry. *The Journal of Risk and Insurance*, 64: 371–382.

Cortes, C., and Vapnik, V. 1995. Support-vector networks. *Machine Learning*, 20: 273–297.

Firth, D. 1993. Bias reduction of maximum likelihood estimates. *Biometrics*, 80 (1): 27–38.

Kunreuther, H., Meszaros, J., Hogarth, R. M., Spranca, M. 1995. Ambiguity and underwriter decision processes. *Journal of Economic Behavior & Organization*, 26: 337–352.

Randall, E. D. 1994. *Introduction to Underwriting*. Insurance Institute of America.

11

Insurance Industry: Claims Modeling and Prediction

11.1 Introduction

Claim management requires applying statistical techniques in the analysis and interpretation of the claims data. The central piece of claim management is modeling and prediction. Two strategies are commonly used by insurers to analyze claims: the two-part approach that decomposes claims cost into frequency and severity components, and the pure premium approach that uses the Tweedie distribution.

In this chapter, we provide a general framework to look into the process of modeling and prediction of claims using the Cox hazard model. The Cox hazard model is a standard tool in survival analysis for studying the dependence of a hazard rate on covariates and time. Although the Cox hazard model is very popular in statistics, in practice data to be analyzed often fails to hold assumptions underlying the Cox model. We use a Bayesian machine learning approach to survival analysis to deal with violations of assumptions of the Cox hazard model.

The term "survival data" has been used in a wide meaning for data involving time to a certain event. This event may be the appearance of a tumor, the development of some disease, cessation of smoking, etc. Applications of the statistical methods for survival data analysis have been extended beyond the biomedical field and used in areas of reliability engineering (lifetime of electronic devices, components, or systems), criminology (felons' time to parole), sociology (duration of the first marriage), insurance (workers' compensation claims), etc. Depending on the area of application, different terms are used: survival analysis—in biological science, reliability analysis—in engineering, duration analysis—in social science, and time-to-event analysis—in insurance. Further, in this chapter, we use terms that are more often used in insurance.

A central quantity in survival (time-to-event) analysis is the hazard function. The most common approach to model covariate effects on time-to-event is the Cox hazard model developed and introduced by Cox (1972). There are several important assumptions that need to be assessed before the model results can be safely applied (Lee, 1992). First, the proportional hazard assumption means that hazard functions are proportional over time. Second, the explanatory variable acts directly on the baseline hazard function and remains constant over time. Although the Cox hazard model is very popular in statistics, in practice data to be analyzed often fails to hold assumptions. For example, when a cause of claims interacts with time, the proportional hazard assumption fails. Or, when the hazard ratio changes over time, the proportional hazard assumption is violated. We present the application of Bayesian approach to survival (time-to-event) analysis that allows dealing with violations of assumptions of the Cox hazard model, thus assuring that model results can be trusted.

This chapter describes a case study intended to indicate possible applications to workers' compensation insurance, particularly the occurrence of claims. We study workers' compensation claims for a period of 2 years. Claims data was provided by a worker compensation insurer that writes approximately $900 million of direct premium annually on a countrywide basis. The risk of occurrence of claims is studied, modeled, and predicted for different industries within several US states.

11.2　Data

The present case study is based on the following policy and claims data (only claims that led to payments are included):

- The start and the end date of the policy;
- Industry in which policy was issued;
- Date of claim occurrence;
- Date of a claim reported;
- State where a claim was reported.

11.3　The Cox Model for Claims Event Analysis

Survival (or time-to-event) function $S(t)$ describes the proportion of policies "surviving" without a claim to or beyond a given time (in days):

$$S(t) = P(T > t)$$

where:

T—survival time of a randomly selected policy

t—a specific point in time.

Hazard function $h(t)$ describes instantaneous claims rate at time t:

$$h(t) = \lim_{\Delta t \to 0} \frac{P\left(t \leq T < t + \Delta t \mid T \geq t\right)}{\Delta t} \tag{11.1}$$

In other words, hazard function $h(t)$ at a time t specifies an instantaneous rate at which a claim happens, given that it had not happened up to time t. The hazard function is usually more informative about the underlying mechanism of claims than survival function.

Cox (1972) proposed a model that does not require the assumption that times of events follow a certain probability distribution. As a consequence, the Cox model is considerably robust.

The Cox hazard model can be written as:

$$h_i(t) = h_0(t) \exp \sum_{j=1}^{k} \beta_j x_{ij} \tag{11.2}$$

where:

$h_i(t)$—the hazard function for subject i at time t

$h_0(t)$—the baseline hazard function, that is, the hazard function for the subject whose covariates $x_1, \ldots, x_k$ all have values of 0.

The Cox hazard model is also called the proportional hazard model if the hazard for any subject is a fixed hazard ratio (HR) relative to any other subject:

$$\text{HR} = h_i(t)/h_p(t) = \left(h_0(t) \exp \sum_{j=1}^{k} \beta_j x_{ij} \right) \bigg/ \left(h_0(t) \exp \sum_{j=1}^{k} \beta_j x_{pj} \right) \tag{11.3}$$

Baseline hazard $h_0(t)$ cancels out, and HR is constant with respect to time:

$$\text{HR} = \exp \sum_{j=1}^{k} \beta_j \left(x_{ij} - x_{pj} \right) \tag{11.4}$$

Estimated survival (time-to-event) probability at time t can be calculated using an estimated baseline hazard function $h_0(t)$ and estimated β coefficients:

$$S_i(t) = S_0(t)^{\exp \sum_{j=1}^{k} \beta_j x_{ij}}$$

$$S_0(t) = \int_0^t h_0(u)\,du \tag{11.5}$$

where:

$S_i(t)$—the time-to-event function for subject i at time t

$x_1, ..., x_k$—the covariates

$h_0(t)$—the baseline hazard function, that is, the hazard function for the subject whose covariates $x_1, ..., x_k$ all have values of 0

$S_0(t)$—the baseline survival function, that is, the survival function for the subject whose covariates $x_1, ..., x_k$ all have values of 0

$\beta_1, ..., \beta_k$—the coefficients of the Cox model.

11.4 Application of the Cox Model for Claims Analysis

We identify three main goals of time-to-event analysis for workers' compensation claims:

1. Estimate survival (time-to-event) function $S(t)$
2. Estimate effects β of industry covariate $x_1, ..., x_k$
3. Compare survival (time-to-event) functions for different industries.

In order to build an appropriate model, we have to address the nature of the claims process. In contrast to biomedical applications where an event of interest, for example, is death and thus can happen only once, in workers' compensation insurance claims happen multiple times, because for each policy there are possible multiple claims. There are many different models that one can use to model repeated events in a time-to-event analysis. The choice depends on the data to be analyzed and the research questions to be answered.

A possible approach is to treat each claim as a distinct observation, but in this case, we have to consider the dependence of multiple claims that belong to the same policy. The dependence might arise from unobserved heterogeneity. Using some simple ad hoc ways to detect dependence (Allison, 2012), we come to the conclusion that the dependence among time-to-event intervals of claims that belong to the same policy is so small that it has a negligible effect on the estimates of the model. Thus, we consider each claim as a single event and can build models that do not account for claims dependence within the same policy.

Below is a short review of different models.

Counting process model

In the counting process model, each event is assumed to be independent, and a subject contributes to the risk set for an event as long as the subject is under observation at the time the event occurs.

The data for each subject with multiple events is described as data for multiple subjects where each has delayed entry and is followed until the next event. This model ignores the order of the events, leaving each subject to be at risk for any event as long as it is still under observation at the time of the event. This model does not fit our application needs because the entry time is considered as a time of the previous event, and time-to-event is calculated as time between consecutive events.

Conditional model I

This conditional model assumes that it is not possible to be at risk for a subsequent event without having experienced the previous event (i.e., a subject cannot be at risk for event 2 without having experienced event 1). In this model, the time interval of a subsequent event starts at the end of the time interval for the previous event. This model does not fit our application needs because it introduces a dependency between consecutive claims.

Conditional model II

This model only differs from the previous model in the way the time intervals are structured. In this model, each time interval starts at zero and ends at the length of time until the next event. This model does not fit our application because it introduces a dependency between claims within the same policy.

Marginal model

In the marginal model, each event is considered as a separate process. The time for each event starts at the beginning of follow-up time for each subject. Furthermore, each subject is considered to be at risk for all events, regardless of how many events each subject actually experienced. Thus, the marginal model considers each event separately and models all the available data for the specific event. This model fits our application needs and is used for the analysis.

11.4.1 Data Transformation

We analyze workers' compensation claims data for the 2-year period, so-called observation period. Each claim is associated with an industry to which employer belongs, and with a state where the accident happened. For example, an employer that belongs to the Entertainment industry with headquarters in NY State may have company offices in different other states, where accidents happen. To prepare this data for the marginal model, each claim event is considered as a separate process. The time to each event is calculated starting from the beginning of the observation period or from the beginning of the policy, whichever happens later. If there are no claim events for a policy during the observation period, the policy is censored at the end of the observation or at the end of the policy, whichever happens earlier.

To note, a subject is said to be censored (Censor = 0) if a policy expired or canceled, or if a claim event did not happen during the observation period.

An example of data prepared for marginal model is presented in Figure 11.1:

1. Policy A starts before January; there are two claims that happened in May and June; the policy ends in August.
2. Policy B starts in March; there is one claim in August; the policy is canceled in October.
3. Policy C starts in April; there are no claims in the observed period of time.

For this example, data is presented as shown in Table 11.1.

In this case study, we present an analysis and modeling performed on claims data for Illinois that contained claims for Consulting, Entertainment, Finance, Hospitality, Manufacturing, Retail, and Utilities industries.

In our analysis, we assume that each claim event is independent within the policy and the industry. For example, if two claims are covered by the same policy, we consider these claims independent. As well, if two claims are covered by different policies, we assume that the claims are independent

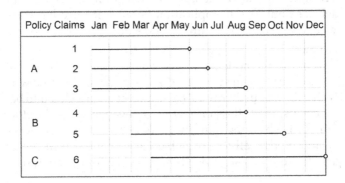

FIGURE 11.1
Claims data presentation.

TABLE 11.1

Claims Data

Policy	Claim	Time-to-Event	Event	Censor
A	1	5	1	1
A	2	6	2	1
A	3	8	3	0
B	4	6	1	1
B	5	8	2	0
C	6	9	1	0

and that the policies have no effect on risk. The data for each policy with multiple claim events is described as multiple claims, where each claim has an entry time at the beginning of the policy or the beginning of the observation period—whichever is later.

11.4.2 Cox Model Assumption Validation

In most insurance risk papers, the authors take the proportional hazard assumption for granted and make no attempts to check that it has not been violated in their data. However, it is a strong assumption indeed. Note that, when used inappropriately, statistical models may give rise to misleading conclusions. Therefore, it is highly important to check underlying assumptions (Arjas, 1988, Gill and Schumacher, 1987). Perhaps the easiest and most commonly used graphical method for checking proportional hazard is the so-called "log-negative-log" plot. For this method, one should plot $\ln(-\ln(S_i(t)))$ vs. $\ln(t)$ and look for parallelism—the constant distance between curves over time. This can be done only for categorical covariates. If the curves show a non-parallel pattern, then the assumption of proportional hazard is violated, and, as a result, the analytical estimation of β coefficients is incorrect.

For claims in seven industries in Illinois, log-negative-log plot is presented in Figure 11.2. This plot shows that the proportional hazard model assumption does not hold: the lines of the log-negative-log plot are not parallel and intersect.

We use industry as a categorical covariate, assuming that time-to-event (survival) functions vary by industries. It is wrong to assume that there is no impact on the baseline hazard function for different values of this covariate variable. For example, hazard changes for Agriculture depending on seasons; or for Transportation depending on weather; or for Hospitality depending on school break schedule.

All these conditions latently depend on time, which means that the impact of industry categorical variable does not remain constant over time, thus violating assumptions of the Cox model. In order to account for season dependency, we introduce time-dependent covariate for the winter season and use an extended Cox model (Hosmer and Lemeshow, 1999):

$$h_i(t) = h_0(t)\exp\left(\sum_{i=1}^{k}\beta_j x_{ij} + \sum_{n=1}^{m}\gamma_n x_{in} g_n(t)\right) \qquad (11.6)$$

where:
$h_i(t)$—the hazard function for subject i at time t
$x_1, \ldots, x_k$—the covariates
$h_0(t)$—the baseline hazard function, that is, the hazard function for the subject whose covariates $x_1, \ldots, x_k$ all have values of 0
$g_n(t)$—the function of time (time itself, log time, etc.)
$\beta_1, \ldots, \beta_k$—the coefficients of the Cox model.

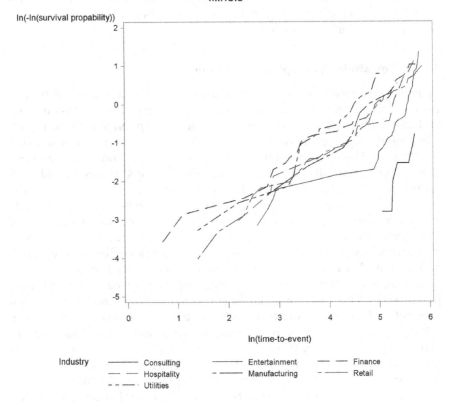

FIGURE 11.2
Log-negative-log plot for Illinois.

Applying this approach to the case of Illinois, our model looks like:

$$h_i(t) = h_0(t) \exp\left(\sum_{j=1}^{6} \beta_j x_j + \gamma \times \text{season} \times \ln(t) \right) \qquad (11.7)$$

where:
  $h_i(t)$—the hazard function for industry $i = 1, \ldots, 6$ at time t, where industries are Consulting, Entertainment, Finance, Hospitality, Manufacturing, and Retail
  $h_0(t)$—the baseline hazard function, in our case the hazard function for one selected industry, by default the last alphabetically ordered industry—Utilities.

$$x_j = \begin{cases} 1, & \text{if } i = j \\ 0, & \text{if } i \neq j \end{cases}$$

$$\text{season} = \begin{cases} 1, & \text{if the claim happened during winter season (months } 11, 12, 1, 2, 3) \\ 0, & \text{if the claim happened during non-winter season (months } 4-10) \end{cases}$$

As there is no reason to prefer any specific industry for a baseline, we choose the last alphabetically ordered industry—Utilities. The selection of Utilities industry as a baseline for hazard means that hazards for all other industries are estimated relative to that of Utilities industry.

Calculation of time-to-event (survival) functions when we have time-varying covariates becomes more complicated, because we need to specify a path or trajectory for each variable (Rodriguez, 2007). For example, if a policy started on the 1st of April, the survival function should be calculated using hazard corresponding to season $= 0$ for time-to-event $t \leq 214$ days (from the 1st of April till the 1st of November), while for time-to-event $t > 214$, it should be calculated using hazard corresponding to season $= 1$. For another example, if a policy started on the 1st of August, the survival function should be calculated using hazard corresponding to season $= 0$ for time-to-event $t \leq 92$ days (from the 1st of August till the 1st of November) and $t > 224$ days (from the 1st of April till the 31st of July), while for time-to-event $92 < t \leq 244$, it should be calculated using hazard corresponding to season $= 1$.

Unfortunately, the simplicity of calculation of $S_i(t)$ is lost: we can no longer simply raise the baseline survival function to a power. For our model, we develop an appropriate formula for calculation of $S_i(t)$:

$$S_i(t) = \left(\frac{S_0\left(t | t < t_1\right) S_0\left(t | t < t_2\right)}{S_0\left(t | t_1 \leq t \leq t_2\right)} \right)^{\exp\left(\sum_{j=1}^{k} \beta_j x_{ij} \right)}$$

$$\times \exp\left(-\exp\left(\sum_{j=1}^{k} \beta_j x_{ij} \right) \times \int_{t_1}^{t | t_1 \leq t \leq t_2} h_0(u) u^{\gamma} \, du \right) \qquad (11.8)$$

where:

t_1—the start day of the winter season relative to the beginning of the policy
t_2—the end day of the winter season relative to the beginning of the policy.

The additional challenge in our data is the reliability of dates related to claims. There are two dates available—the date of the accident caused the claim and the date when the claim was reported. The wide variability of time intervals between these two dates creates an additional challenge in the application of Cox hazard model, as time-to-event becomes an essentially

random variable. To address these problems, as well as assumption violations, we use a Bayesian machine learning approach to estimate coefficients
of the extended Cox hazard model (Ibrahim et al, 2005).

11.4.3 Bayesian Machine Learning Approach

The Bayesian approach is based on a solid theoretical framework. The valid-
ity and application of the Bayesian approach do not rely on the proportional
hazard assumption of the Cox model; thus, generalizing the method to other
time-to-event models and incorporating a variety of techniques in Bayesian
inference and diagnostics are straightforward. In addition, inference does
not rely on large sample approximation theory and can be used for small
samples. In addition, information from prior research studies, if available,
can be readily incorporated into the analysis as prior probabilities. Although
choosing prior distribution is difficult, the non-informative uniform prior
probability is proved to lead to proper posterior probability (Gelfand and
Mallick, 1994). Instead of using partial maximum-likelihood estimation in
the Cox hazard model, the Bayesian method uses the Markov chain Monte
Carlo method to generate posterior distribution by the Gibbs sampler: sample from a specified prior probability distribution so that the Markov chain
converges to the desired proper posterior distribution. However, a known
disadvantage of this method is that it is computation-intensive.

11.4.4 Deployment with SAS

To estimate coefficients of the Cox hazard model, we use SAS® software, spe-
cifically the PHREG procedure, which performs analysis of survival data.
The estimation of the Cox hazard model using the Bayesian approach by SAS
PROC PHREG is implemented in the following way:

```
proc phreg data= CLAIMS_DATA_IL;
   class CLIENT_INDUSTRY;
   model TIME_TO_EVENT*CENSOR(0) = CLIENT_INDUSTRY
   SEASON_EVENT;
   SEASON_EVENT = SEASON*log(TIME_TO_EVENT);
   bayes seed = 1 outpost = POST;
run;
```

CLAIMS_DATA_IL is a SAS dataset that contains data for the state of Illinois
such as industries and time intervals from the beginning of policies to the
date of claims. The sample of rows from CLAIMS_DATA_IL is presented in
Table 11.2.

The CLIENT_INDUSTRY column contains names of industries to which
claims are related. The TIME_TO_EVENT column contains the number of days
to an event calculated starting from the beginning of the observation period

TABLE 11.2

Selected Rows from the CLAIMS_DATA_IL Dataset

CLIENT_INDUSTRY	TIME_TO_EVENT	CENSOR	SEASON
Consulting	119	0	0
Consulting	162	0	0
Consulting	220	1	0
Retail	365	0	0
Retail	237	1	0
Transportation	95	1	0
Transportation	108	1	1
Utilities	7	1	0

or from the beginning of the policy, whichever happens later. The CENSOR column indicates whether the event is a claim (CENSOR = 1), or whether the event is the end of a policy (CENSOR = 0). The SEASON column indicates whether the event happened during the winter season (SEASON = 1) or not (SEASON = 0).

There are two covariates in the model: CLIENT_INDUSTRY and SEASON_EVENT. CLIENT_INDUSTRY is a categorical variable, so it is defined as the covariate in CLASS statement and in MODEL statement. SEASON_EVENT is the time-dependent covariate that represents the following component of the model: season $\times \ln(t)$. SEASON_EVENT is defined in MODEL statement and in the expression that follows the MODEL statement:

```
class CLIENT_INDUSTRY;
model TIME_TO_EVENT*CENSOR(0) = CLIENT_INDUSTRY SEASON_EVENT;
SEASON_EVENT = SEASON*log(TIME_TO_EVENT);
```

The BAYES statement requests a Bayesian analysis of the model by using Gibbs sampling. In the BAYES statement, we specify a seed value as a constant to reproduce identical Markov chains for the same input data. We did not specify the prior distribution, thus applying uniform non-informative prior.

The described PHREG procedure produces an estimation of β and γ coefficients.

However, PROC PHREG does not produce baseline survival function $S_0(t)$ when time-dependent covariate is defined. To calculate the baseline survival function, we use the following workaround (Thomas and Reyes, 2014):

```
data DS;
   set CLAIMS_DATA_IL;
   SEASON_EVENT = SEASON*log(TIME_TO_EVENT);
run;
```

```
data INDUSTRY;
   CLIENT_INDUSTRY = "Utilities";
   SEASON_EVENT = 0;
run;

proc phreg data=DS;
   class CLIENT_INDUSTRY;
   model TIME_TO_EVENT *censor(0) = CLIENT_INDUSTRY
   SEASON_EVENT;
   bayes seed=1;
   baseline out = BASELINE survival = S covariates = INDUSTRY;
run;
```

This step produces baseline survival function $S_0(t)$.

11.4.5 Interpretation of Results

Estimations of β coefficients of the Cox model for all industries except Utilities are presented in Table 11.3. Because the Utilities industry is used as a baseline for hazard, β coefficient for Utilities is equal to 0. Table 11.3 also contains γ coefficient for the SEASON_EVENT covariate.

For the purposes of comparing the risk of claims for different industries, we build survival functions for each industry, and season = 0 (Figure 11.3). According to the survival function for the Utilities industry, for example, there is a 58% chance that there will be no claims before the 100th day of policy, and there is a 1% chance that there will be no claims at all for one-year policy.

The survival functions allow to estimate and to compare the risk of claims among industries. For example, for the Entertainment industry there is a 72% chance that there will be no claims before the 100th day of policy, and a 6.5% chance that there will be no claims at all for a one-year policy. In other words, the Entertainment industry in Illinois presents a 5.5% higher chance than the Utilities industry to have no claims during a one-year policy. Also, we can observe that Entertainment, Manufacturing, and Retail have very similar risks of claims in Illinois. In addition, there is strong evidence that the Consulting industry has a significantly lower risk than the other industries.

TABLE 11.3

Estimations of the Model Coefficients

Industry	Mean Estimate of β	Industry	Mean Estimate of β
Consulting	−2.282	Manufacturing	−0.471
Entertainment	−0.508	Retail	−0.523
Finance	−0.217	Utilities	0.000
Hospitality	−0.145	SEASON_EVENT	0.277

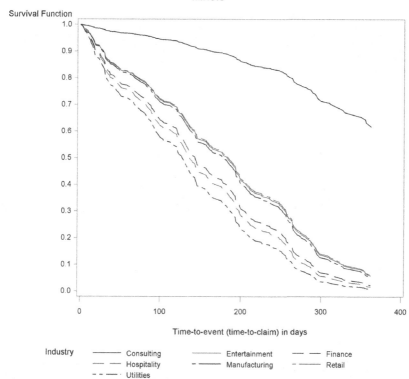

FIGURE 11.3
Survival functions for industries in Illinois.

The hazard function presented in Figure 11.4 shows that the instantaneous claims rate continuously increases, achieving the highest claims rate around the 280th day of policy, and then slightly decreasing. We can also observe that the Consulting industry has somewhat constant and relatively low claims rate through the duration of a policy. The hazard function in Figure 11.4 (as well as in Figure 11.6 and Figure 11.8) was produced with the SMOOTH SAS macro program (Allison, 2012).

The time-dependent covariate SEASON_EVENT is significant with $\gamma = 0.281$. This means that hazard ratio during the winter season in Illinois is 32% higher, controlling for the other covariates:

$$\exp(0.281) - 1 \approx 0.32 = 32\%$$

Estimation of survival (time-to-event) function for a specific policy should take into consideration when the policy started—and thus, when during this policy chances of claims increase due to the winter season.

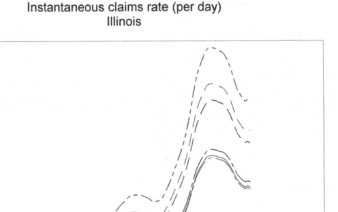

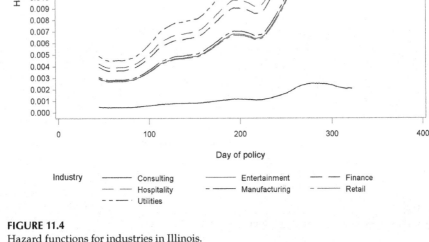

FIGURE 11.4
Hazard functions for industries in Illinois.

Calculation of survival functions when we have time-varying covariates is not straightforward, because we need to specify exactly when a specific policy started and when, relatively to the start date of the policy, the winter season occurred. A proprietary computer program is developed by the authors to calculate $S_i(t)$ for each industry with the time-dependent covariate.

Below we compare two examples mentioned earlier: the case when the policy started on the 1st of April and the case when the policy started on the 1st of August.

If a policy started on the 1st of April, then during time $t \leq 214$ days (from the 1st of April till the 1st of November), *season* = 0. Then, for the duration of time $t > 214$ days till the end of the policy, season = 1. Thus, the survival function is calculated using hazard corresponding to season = 0 for time-to-event $t \leq 214$ days, and for time-to-event $t > 214$ days, it is calculated using hazard corresponding to season = 1. The survival function for this case is presented in Figure 11.5.

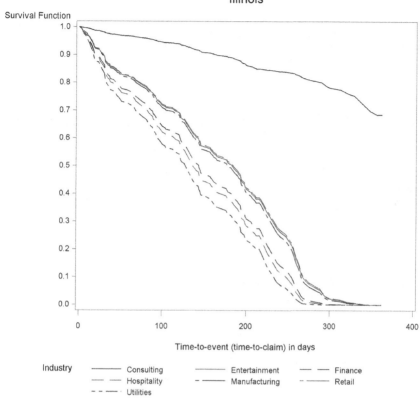

Proportion of policies 'surviving' without a claim beyond a given time (in days)
Illinois

Time-to-event (time-to-claim) in days

Industry	Consulting	Entertainment	Finance
	Hospitality	Manufacturing	Retail
	Utilities		

FIGURE 11.5
Survival functions of industries in Illinois for policies starting on April 1.

In comparison with Figure 11.3, where the winter season was not taken into consideration, we can see that the proportion of survival drops starting from the 214th day of the policy.

Both the Entertainment and Utilities industries have a 0% chance that there will be no claims at all for a one-year policy when we take the winter season into consideration.

In fact, for the Utilities industry there is a 0% chance that there will be no claims even before the 270th day of the policy. However, the chance that Entertainment will "survive" without claims by the 270th day is about 9%.

The hazard function presented in Figure 11.6 shows that the instantaneous hazard of claims sharply increases after $t > 214$ days, achieving the highest claims rate around the 280th day of a policy term.

For the second example, if a policy started on the 1st of August, then during time $t \leq 92$ days (from the 1st of August till the 1st of November) and $t > 244$ days (from the 1st of April till the 31st of July), season = 0. Then, for

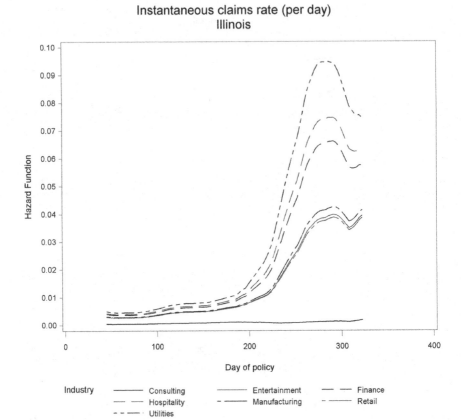

FIGURE 11.6
Hazard functions for industries in Illinois for policies starting on April 1.

the duration of time $92 < t \leq 244$ days of the policy, season = 1. Thus, the survival function is calculated using hazard corresponding to season = 0 for time-to-event $t \leq 92$ and $t > 244$ days, and for time-to-event $92 < t \leq 244$, it is calculated using hazard corresponding to season = 1. The survival function for this case is presented in Figure 11.7.

In comparison with Figure 11.3, where the winter season was not taken into consideration, we can see that the proportion of survival drops before the 100th day of the policy.

For the Utilities industry, there is a 44% chance that there will be no claims before the 100th day of a policy accounting for winter season vs. 65% without accounting for the winter season. After that, the chances are dropping, and by the 210th day of the policy, there is a 0% chance that there will be no claims in the Utilities industry, accounting for the winter season.

For the Entertainment industry, there is a 66% chance that there will be no claims by the 100th day of a policy when we take winter season into

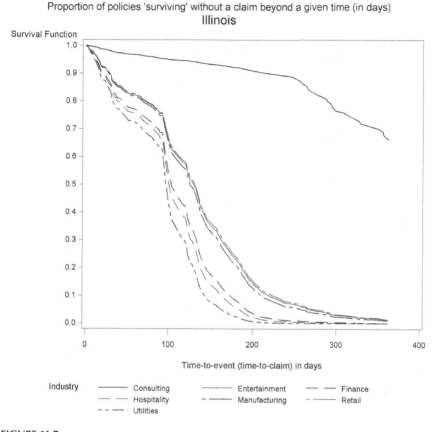

FIGURE 11.7

Survival functions for industries in Illinois for policies starting on August 1.

consideration—vs. 72% otherwise. Also, there is about a 1% chance that there will be no claims at all for a one-year policy—in comparison with a 6.5% chance when we do not take the winter season into consideration.

The information revealed by the presented models can be used for the purposes of underwriting and pricing, for the development of new insurance products, and for marketing. For example, insurers can estimate the risk of claims more accurately depending not only on the industry, but also on the time period when the policy is started. Insurers can better manage anticipation of losses related to claims. In addition, insurers can develop new workers' compensation products for a duration shorter than 1 year. In this case, the insurance during the periods of lower risk will have a lower premium and therefore higher acceptance rate by customers. Referring to the example when a policy starts on the 1st of April, the 6-month policy would have a significantly lower risk and will justify lower premiums. The marketing of such new products will attract companies seeking workers' compensation.

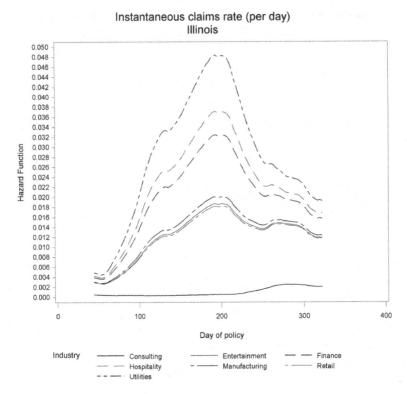

FIGURE 11.8
Hazard functions for industries in Illinois for policies staring on August 1.

11.5 Summary

An ultimate goal of insurance risk assessment is to create a profitable portfolio and to fit the right price to the right risk. This complex problem comprises from multiple parts, including estimation of risk, estimation of price, and monitoring of market changes. In this chapter, we discussed one part of this complex problem—estimation of risk of workers' compensation claims for different industries and states with season-dependent factor. Our method to estimate hazard function using a Bayesian machine learning approach allows estimating the risk of claims per industry and state, ranking industries by risk within states, and estimating the risk depending on time-varying covariates like a season. As a next step to build a profitable portfolio, the severity of claims should be included in the analysis, which eventually will allow re-evaluating premiums and insurance products to increase the profitability of portfolios.

References

Allison, P. D. 2012. *Survival Analysis Using SAS*. Cary, NC: SAS Publication.

Arjas, E. 1988. A graphical method for assessing goodness of fit in Cox's proportional hazards model. *American Statistical Association*, 83: 204–212.

Cox, D. R. 1972. Regression models and life-tables (with discussion). *Journal of the Royal Statistical Society – Series B*, 34: 187–220.

Gelfand, A. E., and Mallick, B. K. 1994. *Bayesian analysis of semiparametric proportional hazards models*. Technical Report No. 479. Department of Statistics, Stanford University.

Gill, R., and Schumacher, M. 1987. A simple test of the proportional hazards assumption. *Biometrika*, 74: 289–300.

Hosmer, D. W., and Lemeshow, S. 1999. *Regression Modeling of Time to Event Data*. New York: John Wiley & Sons, Inc.

Ibrahim, J. G., Chen, M. H., and Sinha, D. 2005. *Bayesian Survival Analysis*. Hoboken, NJ: Wiley Online Library.

Lee, E. T. 1992. *Statistical Methods for Survival Data Analysis*. 2nd Ed. Oklahoma City: John Wiley & Sons, Inc.

Thomas, L., and Reyes, E. M. 2014. Tutorial: Survival estimation for Cox regression models with time-varying coefficients using SAS and R. *Journal of Statistical Software*, 2014: 61.

References

Index

Note: **Bold** page numbers refer to tables and *italic* page numbers refer to figures.

Printed in the United States
by Baker & Taylor Publisher Services